A Colour Atlas of
CUCURBIT DISEASES

<u>Acknowledgements</u>

A book is the fruit both of its authors' labours and of the constructive criticism of those who read the manuscript.

We would like to thank all those who helped, and in particular, Mr Messiaen and Dr Fletcher who very kindly agreed to write forewords to the French and English editions.

We also wish to thank the following for their invaluable contribution: Ms Arpin, Ms Cazenave, Ms Glavier, Ms Lecoq, Ms Médalin, Ms Remaud, Ms Pichot, and Messers Jailloux, Lafon, Larignon, Laterrot, Mas, Nicot, Rocamora and Rougier.

A Colour Atlas of

CUCURBIT DISEASES

Observation, Identification & Control

D. Blancard
INRA Vegetable Pathology Unit
Villenave-d'Ornon, France

H. Lecoq
INRA Vegetable Pathology Unit
Montfavet, France

M. Pitrat
INRA Plant Improvement Unit
Montfavet, France

with the assistance of
M. Javoy
Loiret Chamber of Agriculture
Orleans, France

Foreword to the French edition
C.M. Messiaen

English translation by **Jean Kirby**

Scientific advice and a Foreword to the English edition
John T. Fletcher
ADAS, Wye, England

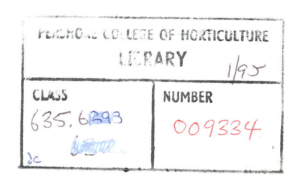

MANSON
PUBLISHING
LONDON

Halstead Press: an imprint of
John Wiley & Sons
NEW YORK   TORONTO

Copyright © Manson Publishing Ltd, 1994
Published by Manson Publishing Ltd, 1994
Printed by Centre Impression, Limoges, France
ISBN 1-874545-15-4
Wiley ISBN 0-470-23416-4

First published in French as *Maladies des Cucurbitacées*
by INRA, Route de Saint Cyr, 78026 Versailles and P.H.M.–
Revue Horticole, B.P. 1516, 87021 Limoges.
French edition © INRA and P.H.M.–Revue Horticole 1991.
ISBN 2–7380–0311–7

This edition is published with the help of the French Ministry of Culture.

For full details of all Manson titles please write to
Manson Publishing Ltd, 73 Corringham Road,
London NW11 7DL, England.

Published in North and South America by John Wiley & Sons Inc.,
605 Third Avenue, New York, NY 10158-0012

Library of Congress Cataloging-in-Publication Data
Blancard, Dominique.
    [Maladies des cucurbitacées. English]
  A colour atlas of cucurbit diseases: observation, identification
& control / D. Blancard, H. Lecoq, M. Pitrat ; with the assistance
of M. Javoy ; English translation by Jean Kirby ; scientific advice
and a foreword to the English edition, John T. Fletcher.
    p.   cm.
  Includes bibliographical references (p. 299) and index.
  ISBN 0–470–23416–4
  1. Cucurbitaceae — Diseases and pests — Identification.
2. Cucurbitaceae — Diseases and pests — Atlases. 3. Cucurbitaceae —
Diseases and pests — Control. I. Lecoq, H. II. Pitrat, M.
SB608.C885.B5813 1994
635'.61—dc20   94–20419
                                                              CIP

Contents

Foreword to the French Edition

The older market gardeners of Provence remember the 60s as a golden age, a time when their melon plants grew so vigorously that 'they had to cut down the foliage to find the fruit'.

Today things are not so easy. The melon crop has been affected by fusarium wilt, mildew, and mosaic diseases; as soon as glasshouse cultivation of cucumbers began in France, the crop fell prey to *all* the classic diseases known to Dutch and British growers, with the exception of fusarium wilt.

The Plant Pathology and Plant Improvement Units of INRA immediately joined forces in an effort to find pesticidal, or better still, genetic solutions to these problems, particularly where they concerned the melon and the courgette. Plant pathologists at INRA, Dijon, successfully developed melon grafting, and introduced the concept of soils suppressive of fusarium wilt.

In northern Europe, understanding of the diseases which affect cucumbers has continually increased. However, the plant pathologists and geneticists who worked in the 1970s and 1980s, the period with which the authors of this book were concerned, did not find their task easy.

The appearance of mildew, the reappearance of gummosis, the threatening and worldwide acceleration in viral diseases all called for further research. The sheer variety of symptoms which the grower can encounter is motive enough for compiling this beautifully illustrated 300-page book. It includes the necessary diagnostic aids for recognising both the 'textbook symptom', rarely seen in all its splendour, and also the symptom as it more often appears in life, with its many ambiguities.

'Non-parasitic symptoms', appearing in increasing numbers as cultivation methods diversify, are covered in detail.

The grower should not be discouraged at the sight of all the misfortunes which await him. This book will show how to prevent a warning from turning into disaster, or if the worst has happened, how to avoid a recurrence.

C.M. Messiaen

Foreword to the English Edition

Cucurbits are members of the family Cucurbitaceae. This is a large group of plants with about 90 genera and over 700 species largely confined to warmer climates. A number of different species are cultivated and are important food and ornamental crops. The climatic range of the horticultural species is extended in cooler countries by cultivation under the protection of glass or plastic. There are few countries where cucurbits are not a common component of the diet. Cucumbers, marrows, squashes, courgettes, gourds, melons, water melons and even luffas are found in shops in cities and towns in most countries. These are worthwhile and profitable crops for commercial growers and amateurs alike. But the successful production of such crops is often hindered by the unexpected development of pests and diseases. Many cucurbits are susceptible to a wide range of fungal, bacterial and viral pathogens. They are also ideal hosts for the propagation of the pests that are efficient virus vectors. Because of their susceptibility epidemics develop very rapidly and there have been recent and devastating examples of these in field grown cucurbits in California. Similarly, in the enclosed environment and often monoculturing of the greenhouse, serious crop loss can occur. Cucumber roots, for example, are very prone to the ravages of the root decaying fungal pathogens and those with experience of this crop often express the view that cucumbers are either in a state of vigorous growth or are dying – there being no inbetween. With high cash value crops such as cucumbers, melons, water melons, etc., and the potential for epidemic loss, it is vitally important to recognise the early signs of disease so that disease development can be prevented. Accurate diagnosis is the first but essential step to successful disease control.

A Colour Atlas of Cucurbit Diseases by D Blancard, H. Lecoq and M. Pitrat covers the diseases and disorders of the most commonly grown cucurbits. Pest damage is also featured. The illustrations of symptoms as found in naturally affected crops are superb. Where pathogens affect different plant parts, i.e. stems, leaves and fruits, all of the relevant symptoms are included. With these excellent illustrations grouped according to plant part, disease diagnosis is greatly simplified. In addition essential information is also included for each of the major diseases allowing a greater understanding of why the disease has developed and how it can be controlled.

This Colour Atlas therefore includes all of the ingredients to enable successful diagnosis and thereby control of cucurbit diseases as well as biological information which is useful in the avoidance of future outbreaks.

Because the cucurbit crop has almost no geographic boundaries, this book will be an invaluable aid for all those involved or interested in growing these crops. It will find a special place in the U.S.A. and Canada, Northern and Mediterranean Europe, South Africa, the Far East – especially Japan – and Australasia, where cucurbits are economically very important.

J.T. Fletcher

Introduction

In a previous volume on tomato diseases we described the steps that a plant pathologist would take when identifying a disease affecting that species. We emphasised the advantages of early diagnosis, while pointing out that this could be difficult in view of the numerous and possibly confusing symptoms.

Encouraged by the success of the tomato book and convinced that both growers and instructors would welcome the availability of diagnostic aids, we have produced a new book which enables the reader to identify the diseases (parasitic and non-parasitic) affecting the main varieties of the cultivated Cucurbitaceae and also provides information on suitable methods of control.

The book is arranged in two parts:

- Part One deals with the diagnostic aspect and is illustrated by 472 colour photographs and several line drawings which simplify inspection of the affected plants. It is designed for easy reference and provides straightforward definitions of symptoms. It is also an educational handbook from which the reader can acquire the skills of observation and symptom recognition essential to the establishment of a reliable diagnosis.

  When using this book, we would urge our readers to follow the procedures described in the following pages. The more experienced will undoubtedly be tempted to take shortcuts. If they do, they may find themselves confused by the organisation of the book as it is not based on the type of pathogen (often micro-organisms), but on the symptoms which characterise these agents.

- In Part Two, for use once identification has been made, there is a factfile provided for each of the pathogens. These files include data on their main biological characteristics and information on suitable control methods. In addition to the measures for immediate application, advice on preventing a recurrence of the disease in following seasons has been given wherever possible.

How to use this book

Before attempting a diagnosis:

1. **Choose good quality specimens,** preferably whole plants, showing typical, early stage symptoms of the disease.

2. **Collect as much information as possible:**

- **On the disease** (spread in the crop and location on plants (see pages 14, 15 and 16), rate of development, climatic conditions at or before it appearance or apparently favouring its spread, etc.);
- **About the crop** (varietal characteristics, seed quality, etc.)
- **In the field** (previous crops, fertilisers or used manure, etc.)
- **About the cultivation operations carried out** (method and frequency of irrigation as well as amount of water used each time, pesticide applications to the crop or in the vicinity, etc.)

Diagnosis:

1. **Locate the symptoms on the diseased plants and define their nature.**

	Reference colours in the book	*Where to find the observation guides*
Plants		14, 15, 16
Leaves		
Deformations		20, 21
Discolorations		38, 39, 40, 42
Spots on leaves		72, 73
Wilting, desiccation (with/without yellowing)		106, 107
Roots, collar, lower stem		106, 107, 114, 115, 116
Stem (exterior, interior)		106, 107, 132, 133, 144
Flowers and fruit		154, 155

10

2. Turn to the pages which deal with the part of the plant which is diseased and which describe the type of symptom observed.

At the beginning of the section you will find:
- **Symptoms examined**
- **Possible causes**

(Many symptoms can be associated with more than one cause.)

In the case of leaf decay the numerous symptoms have been divided in accordance with observed criteria into four subsections:
- **Leaf deformations**
- **Leaf discolorations**
- **Leaf spots**
- **Wilting or desiccation of leaves**

Decay in the roots, collar and stem base has been grouped together in the same section. The micro-organisms responsible very often affect all three areas at the same time, making it difficult to separate them.

3. Select a symptom and turn first to the pages that describe it, then if necessary refer to all the symptoms in that section.

For each symptom one or several **possible causes** are suggested. (There may be several causes associated with one symptom.)

4. Determine the cause of the symptom.

To sort out the likely cause from those described:
- Compare the symptom or symptoms observed in the plants with those illustrated in the numerous photographs;
- Study the **additional diagnostic guidelines**;
- If necessary, refer to the symptoms in the other relevant pages.

Text referring to diseases rare or as yet unknown in Europe is printed in a blue box.

Control

Methods for controlling parasitic micro-organisms will be found in the factfiles which make up Part Two of the book. These are arranged as follows:

- **Symptoms** (numbers of the photographs showing the disease symptoms).
- **Main characteristics of the pathogen** (prevalence and economic importance, source of infection, spread, favourable development conditions, etc.).
- **Control methods** (applicable to the present crop and to the following season's crop).

Where non-parasitic diseases are concerned the methods used to limit their spread are often related to the cause or causes of the infection (these are described in Part One). In many cases, poor control of climatic conditions and/or unsuitable cultural conditions are responsible, i.e. frost damage, phytotoxicity in many forms, root asphyxia, various deficiencies, etc. In these cases the errors in management must be corrected to ensure better growing conditions for the plants.

Three **appendices** complete the book:

Appendix I: Summary of the main problems which affect the health of sproutings (stages: sowing, pricking-out, planting) and the measures which must be taken to avoid them. Included here is information on disinfection of substrates and soils.

Appendix II: Summary of damage caused by the main pests affecting Cucurbitaceae and by dodder.

Appendix III: Botanical data concerning Cucurbitaceae. List of current principal commercial varieties which are disease resistant.

Part One

Diagnosis of Parasitic and Non-parasitic Diseases

OBSERVATION GUIDE

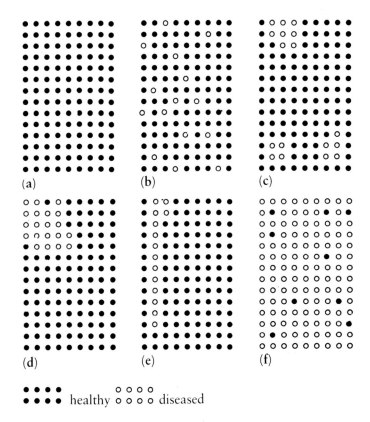

(a)   (b)   (c)

(d)   (e)   (f)

●●●●
●●●● healthy ○○○○ diseased
           ○○○○

Distribution of diseased plants within the crop
(H = healthy   D = diseased)

1   (a) Healthy crop.
     (b) Randomly dispersed diseased plants.
     (c) Several small, dispersed foci of disease.
     (d) Large focus of disease.
     (e) Long or short lines of diseased plants.
     (f) Disease affecting most of crop.

1 Several plants becoming yellow and desiccated in the same part of the field, forming a focus of disease—*Fusarium oxysporum* f.sp. *melonis*.

2 A large area of this crop has been cleared (D), the highly susceptible plants never having developed—Phytotoxicity.

3 Generalised mildew attack on a crop growing under a plastic tunnel—*Pseudoperonospora cubensis*.

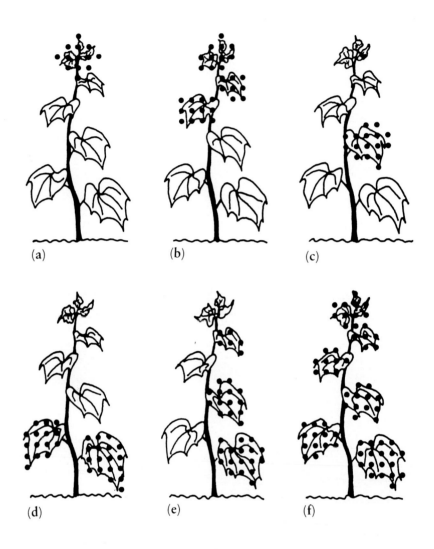

Location of foliar symptom/s on the plant/s examined

(a) Apex–terminal bud
(b) Young leaves (top of plant)
(c) Separate and random
(d) Old leaves (plant base)
(e) Leaves on one side only of the plant (unilateral)
(f) All leaves (generalised)

Abnormalities and Decay in Leaves

Symptoms observed on the leaves have been divided into four sub-sections, for easier reference.

Deformation of leaves

SYMPTOMS EXAMINED

- Dwarfed, stunted vegetation
- Shrivelled leaves
- Wrinkled, curled leaves
- Leaves showing abnormalities of size and shape (serrated, filiform, etc.)

POSSIBLE CAUSES

- Cucumber Mosaic Virus (CMV)
- Squash Mosaic Virus (SqMV)
- Zucchini Yellow Mosaic Virus (ZYMV)
- Water Melon Mosaic Virus type 2 (WMV2)
- Cucumber Toad Skin Virus (CTSV)
- Papaya Ring Spot Virus (PRSV ex WMV1)
- Complex of viruses

- Adverse cultivation methods
- Adverse climatic effects
- Calcium deficiency
- Various phytotoxic effects

- Aphids (see Appendix II)

DIFFICULT DIAGNOSIS

Leaf symptoms are very often similar even though caused by different pathogens and this makes identification difficult. Therefore, we suggest that you consider all the symptoms mentioned in this sub-section and also in the subsection concerned with leaf discoloration. Because these symptoms may be typical of several diseases (except for some which are strongly characteristic), more than one possible cause can often be put forward.

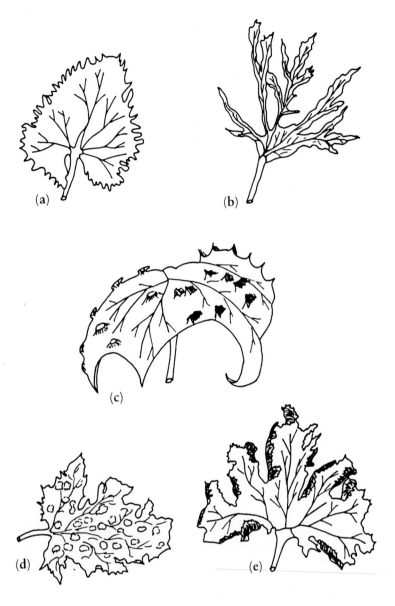

Some leaf deformations

(a) Melon leaf with deeply serrated edges.
(b) Laciniate or filiform courgette leaf.
(c) Shrivelled cucumber leaf, covered on the right side by outgrowths, on the left by pustules.
(d) Melon leaf, showing deformation and pustules.
(e) Courgette leaf, showing rolled edge.

Examples of leaf deformation

4 Shrivelled leaves (CMV).

5 Partially rolled leaf (CMV).

6 Leaves with pustules and serrated edges; fruit slightly mottled (ZYMV).

7 Laciniate leaf (ZYMV).

8 Base of courgette plant with growth stunted from the very early stages after early infection by a virus—Zucchini Yellow Mosaic Virus (ZYMV).

9 The many laciniate leaves on this courgette plant give it a weak, abnormal appearance—Zucchini Yellow Mosaic Virus (ZYMV).

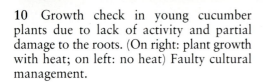

10 Growth check in young cucumber plants due to lack of activity and partial damage to the roots. (On right: plant growth with heat; on left: no heat) Faulty cultural management.

11 Apex of cucumber plant forming a rosette; stem growth check is a result of a viral infection—Cucumber Mosaic Virus (CMV).

Dwarfed, stunted vegetation

The leaf deformations caused by the diseases described in this section very often give a characteristic appearance (temporary or permanent) to either the whole plant or just its apex, distinguishing it from healthy plants.

When young plants are affected in the early stages their growth may be permanently stunted, or it may be greatly retarded resulting in a dwarfed appearance.

If infection occurs at a later stage, only the apex and the new shoots will be affected and show the symptoms described below.

Certain pathogenic fungi, air-borne or soil-borne, and sometimes faulty cultural management, may cause similar symptoms. Very often they cause other, more characteristic changes which should enable you to distinguish them from the diseases dealt with in this section.

12 The very short internodes indicate a growth check in this melon plant—Zucchini Yellow Mosaic Virus (ZYMV).

13 Cucumber leaf shrivelled towards the base, with chlorotic patches—Cucumber Mosaic Virus (CMV).

14 Slightly withered courgette leaves with star-shaped yellow patches—Cucumber Mosaic Virus (CMV).

15 Shrivelled, reticulate courgette leaf, with vein yellowing—Cucumber Toad Skin Virus (CTSV).

16 Shrivelled cucumber leaf, slightly chlorotic round edge—adverse climatic conditions.

17 Cucumber plants showing retarded growth, very short internodes and chlorotic leaves with a tendency to curl—calcium deficiency.

24

SHRIVELLED LEAVES

POSSIBLE CAUSES

- **Cucumber Mosaic Virus (CMV)** (factfile 29)
- **Cucumber Toad Skin Virus (CTSV)** (factfile 34)

- **Aphids** (see Appendix II)

- **Various phytotoxicities**
- **Faulty cultural management**
- **Adverse climatic conditions**
- **Calcium deficiency** (see deficiencies pp. 60–61)

ADDITIONAL DIAGNOSTIC GUIDELINES

	CMV	CTSV
Susceptible hosts	melon, cucumber, courgette	cucumber
Frequency of attacks	+ + +	+/−
Number of plants affected	Numerous, often when grown under cover	A few at random
Other symptoms characteristic of viruses in cucumbers	Mottling (photo. 13) Wilting of whole plant (photo. 202)	Necrosis of first leaves infected. Yellowing of veins (photo. 15)

Other viruses (Zucchini Yellow Mosaic Virus, Water Melon Mosaic Virus type 2) may also cause the leaves to shrivel; this symptom is rare but may appear unexpectedly when the disease is already established.

    Faulty cultural management or adverse climatic conditions (calcium deficiencies, excess humidity, etc.) are likely to influence leaf development. Where growth is disturbed and often reduced (see **16–17**), leaf shrivel is commonly seen.

    Certain pesticides when applied too generously or sprayed in unsuitable weather conditions can cause **phytotoxicity**, in particular round the leaf rim where there is a tendency for the pesticide to accumulate. The necrosed tissue at the edge of the lamina causes distortion in the shape of the developing leaf which frequently takes on a shrivelled appearance (see also page 35).

18 Irregular sized blisters scattered over the entire lamina—Papaya Ring Spot Virus (PRSV).

19 Regular and generalised wrinkling over entire lamina—Zucchini Yellow Mosaic Virus (ZYMV) with CMV and WMV2.

20 Leaf partially bubbled and deformed, also showing enations—Zucchini Yellow Mosaic Virus (ZYMV).

21 Highly denticulate leaf covered by pronounced bubbles—Zucchini Yellow Mosaic Virus (ZYMV).

MELON

In general, the bubbles, blisters, etc, are darker than the rest of the lamina.

Refer also to the tables on page 43 which indicate how often and at what season outbreaks of the viruses occur.

WRINKLED, BUBBLED LEAVES

POSSIBLE CAUSES

- **Cucumber Green Mottled Mosaic Virus (CGMMV)** (factfile 2)
- **Cucumber Mosaic Virus (CMV)** (factfile 29)
- **Squash Mosaic Virus (SqMV)** (factfile 27)
- **Zucchini Yellow Mosaic Virus (ZYMV)** (factfile 32)

- **Water Melon Mosaic Virus type 2 (WMV2)** (factfile 31)
- **Cucumber Toad Skin Virus (CTSV)** (factfile 34)
- **Papaya Ring Spot Virus (PRSV, ex WMV1)** (factfile 30)

- **Adverse climatic conditions**

ADDITIONAL DIAGNOSTIC GUIDELINES

(Refer to paragraph 'Discoloration of Leaves' [especially heading 'Mottled leaves'])

Refer also to the tables on page 43 which indicate how often and at what season outbreaks of the viruses occur in France, and to the photographs of virus symptoms in the melon (pages 44–46).

VIRUS	SYMPTOMS on leaves	on fruit
CMV	Growth retardation. Mottling (**45, 53, 54**).	Marbling. Aborted flowers, reduced yield (**389**).
ZYMV	Dwarfing, serious deformation, enations. Prominent, yellowing veins (**58–83**). Lethal wilting (**219**) (depending on pathotype and variety).	Slight external mottling, external cracks. Internal marbling, hardening of flesh (**381–382**).
WMV2	Reduction in leaf surface area. Mottling often causing deformation (**57**). Vein banding.	Pronounced mottling, fruits may be dented (**390**).
PRSV	Deformed leaves have striped appearance (**59**). Generalised yellowing.	Mottling with raised dark green patches.
SqMV	Serious deformation following pronounced vein banding (**60**). Recovery possible.	Reduced yield. Mottling.

22 Cucumber leaf showing slight, irregular flattening and slight bubbling—Cucumber Mosaic Virus (CMV).

24 Deformed leaf with flattened areas, showing discrete marbling. Cucumber Green Mottle Mosaic Virus (CGMMV).

23 Young cucumber leaves with mottling and wrinkling—Zucchini Yellow Mosaic Virus (ZYMV).

25 Crinkled areas on some sectors of the leaves, the rest of the lamina appearing normal—adverse climatic conditions (water condensation).

CUCUMBER

| VIRUS | SYMPTOMS | |
	on leaves	on fruit
CMV	Retarded growth. Wilting (**202**) Mottling with bubbles (**61**)	Marked mottling, fruit surface is slightly 'dented' (**394**)
CGMMV	Lamina has flattened appearance with light [–] dark green mottling (**62**)	No serious deformation Reduced yield.
ZYMV	Mottling causing varying degree of deformation (**63**) Prominent veins.	Marbling (**395**)

Refer also to the tables on page 43 which indicate how often and at what season outbreaks of the viruses occur in France (page 43) and to the photographs of the other symptoms of cucumber virus diseases (page 48).

Adverse climatic conditions

Condensation often forms on the inside of the glass during the night. When it is heavy, cold water may trickle down onto the plants and accumulate on various parts of the leaf. The tissues of the lamina react to this low temperature water by producing crinkled areas (**25**). Normally these distortions have no adverse effect on the plant.

(Virus affecting courgettes: see photographs and tables on pages 32–33)

27 Dark green bubbles contrasting with rest of pale yellow lamina in a courgette leaf—Zucchini Yellow Mosaic Virus (ZYMV).

26 Discrete flattened areas partially covering the lamina of a courgette leaf—Papaya Ring Spot Virus (PRSV).

28

28 Lanceolate leaf with slight mottling—Zucchini Yellow Mosaic Virus (ZYMV) or Squash Mosaic Virus (SqMV).

29

29 Incurving leaf with abnormally serrated edge—Zucchini Yellow Mosaic Virus) (ZYMV) with Water Melon Mosaic Virus type 2 (WMV2).

30

30 Very serrated leaf showing large bubble—Zucchini Yellow Mosaic Virus (ZYMV).

MELON

Virus affecting the melon (see photographs and tables on pages 27–44–45–46).

LEAVES SHOWING ABNORMALITIES OF SIZE AND SHAPE (SERRATED, FILIFORM, ETC)

POSSIBLE CAUSES

- **Cucumber Green Mottle Mosaic Virus (CGMMV)** (factfile 26)
- **Cucumber Mosaic Virus (CMV)** (factfile 29)
- **Squash Mosaic Virus (SqMV)** (factfile 27)
- **Zucchini Yellow Mosaic Virus (ZYMV)** (factfile 32)
- **Papaya Ring Spot Virus (PRSV, ex-WMV1)** (factfile 30)
- **Water Melon Mosaic Virus type 2 (WMV2)** (factfile 31)
- **Complex of viruses**

- **Various phytotoxicities**

ADDITIONAL DIAGNOSTIC GUIDELINES

- **Various viruses**

See tables and photographs in the preceeding and following pages.

31 Branch with several very deformed leaves, bubbled and serrated—Zucchini Yellow Mosaic Virus (ZYMV).

32 Peripheral areas of lamina are lighter in colour and drawn out—Water Melon Mosaic Virus type 2 (WMV2).

33 Slightly filiform leaf, partially incurved and bubbled—Papaya Ring Spot Virus (PRSV).

34 Highly denticulate and stunted leaf with a 'flame' appearance—Zucchini Yellow Mosaic Virus (ZYMV).

35 Filiform or laciniate leaf—Zucchini Yellow Mosaic Virus (ZYMV).

COURGETTE

VIRUS	SYMPTOMS	
	on leaves	on fruit
CMV	Mottling, can cause great deformation (**51**). Wrinkled and rolled leaves near ground (**14**). Mottling and deformation of petioles.	Mottling causing deformation. Fruit has pitted appearance (**398–405**).
ZYMV	Serious foliar deformation (leaves in strips). Prominent veins. Mottling, yellowing. (**50–67**)	Mottled and blistered fruit (**399–405**)
WMV2	Laminar serration is more frayed (**66**) (depending on strain), Marbling.	Green spots on courgettes with yellow fruit. Normally, green fruit has no symptoms.
PRSV	Leaves more/less laciniate with silver discoloration (**68**). Mottling.	Fruit dented, showing mottling (**405**)

See also the tables and photographs and virus symptoms in the courgette (pages 50–51).

A virus similar to the Papaya Ring Spot Virus (PRSV) causes slightly different foliar symptoms (filiform deformation without silver colouring). Normally found in Africa, but has been reported on one occasion in Spain.

36 Young laciniate leaf, the result of accidental contact with glyphosate—Phytotoxicity.

36

37 Malformed young melon leaf, caused by slight burn suffered very early on at meristem stage—Phytotoxicity.

38 Melon leaf with local burns, the result of treatment, giving rise to necrosis and deformation of the lamina—Phytotoxicity.

39 Periphery of lamina is serrated, the veins form more acute angles—Phytotoxicity.

40 Young branch with denticulate and wrinkled leaves—Phytotoxicity.

<mark>MELON</mark>

Some examples of phytotoxicity in the melon. Similar symptoms can be seen in other members of the Cucurbitaceae family.

● Various phytotoxicities

These are quite difficult to determine and very often the grower refuses to admit to any error or negligence on his part.
In most cases, the cause or causes of these phytotoxicities can be confirmed by studying their occurrence in relation to time and locality.

● Occurrence in relation to time

Symptoms can become apparent:
○ Quite rapidly (= immediate cause and effect relationship) after the application of pesticide to a crop or in its vicinity (spray application);
○ Later where previous cultural operations have left an unfortunate legacy (preceding annual or perennial crop cleared with herbicides which have persisted or have not been leached out owing to a dry winter; perinnials cleared with herbicides over several years = accumulated herbicide) or following the application of straw or straw manure from cereal crops.

● Occurrence in relation to location

This can vary according to the source of the phytotoxic compound.
If the phytotoxic compound is applied to the leaves the distribution of diseased plants may be:
○ Evenly spread all over the area;
○ At the beginning of the row;
○ Near the glasshouse doorways or ventilators;
○ On one side of the plants.
If the compound has persisted in the soil as residues, distribution of the diseased plants may be:
○ Spread more or less evenly all over the area;
○ Along the edges of the glasshouses.

Sometimes the fruit may also show symptoms (see **329–330**, page 164).

● **Have you thoroughly washed out your sprayer?** Be careful if you use water from irrigation channels or even wells, because they may be polluted with herbicide.

(See page 62 for other symptoms of phytotoxicity.)
To solve your problem you can adopt the following measures:
○ Carefully identify the source of phytotoxicity;
○ Take steps to ensure that it cannot occur again;
○ Do not destroy the plants immediately, but carry on growing the plants as usual and observe how they develop; this will depend chiefly on the constituents of the compound and the persistence of the product or products in question;
○ There are no other specific measures that we can recommend.

Discoloration of leaves

SYMPTOMS EXAMINED

- Mottled leaves (Mosaic diseases)
- Leaves partially or totally yellow or chlorotic
- Leaves partially or totally white
- Leaves silvery or with a metallic sheen

POSSIBLE CAUSES

- *Fusarium oxysporum* f.sp. *melonis*
- *Pseudoperonospora cubensis*

- Melon luteovirus
- Cucumber Yellow Virus (CYV)
- Musk Melon Yellow Virus (MYV)
- Lettuce Infectious Yellow Virus (LIYV)
- Cucumber Green Mottled Mosaic Virus (CGMMV)
- Cucumber Mosaic Virus (CMV)
- Squash Mosaic Virus (SqMV)
- Zucchini Yellow Mosaic Virus (ZYMV)
- Water Melon Mosaic Virus type 2 (WMV2)
- Cucumber Toad Skin Virus (CTSV)
- Papaya Ring Spot Virus (PRSV ex WMV1)

- Genetic anomalies (chimera, physiological blemishes in courgettes)
- Silvering in courgettes
- Excess salinity
- Hypersusceptibility to mildew
- Physiological yellowing of courgette leaves
- Various phytotoxicities.

- Spider mites (see Appendix II)
- Thrips (see Appendix II)

DIFFICULT DIAGNOSIS

Diseases originating in leaf discolorations often have several symptoms in common (in particular, yellowing of leaves), which makes identification difficult. We suggest that you refer to all the symptoms in this sub-section. In addition, such diseases cause deformation in leaves and it would be advisable to consult that section also.

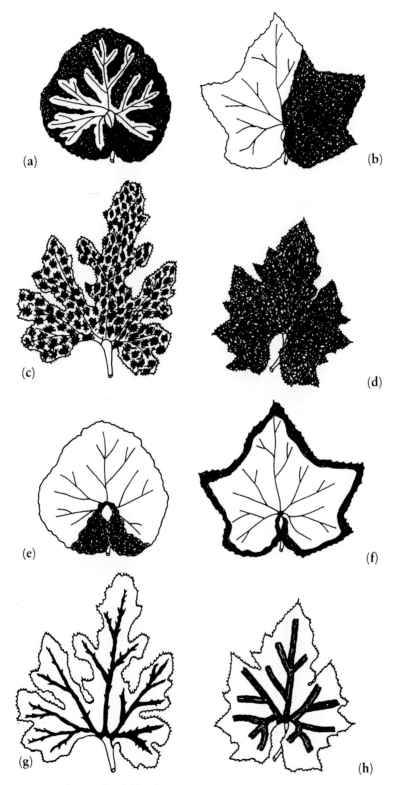

Location and appearance of some leaf discolorations.

(a) Between the leaf veins.
(b) On one side only of the leaf (unilateral).
(c) Spots with diffused edges over entire leaf surface.
(d) Covering the entire leaf surface evenly.
(e) At the leaf base.
(f) Round the periphery of the leaf.
(g) Spreading along the leaf veins.
(h) Surrounding the veins (Vein banding).

Examples of discoloration in leaves

41 Mottled leaf (ZYMV).

42 Leaves partially affected with chlorosis (Phytotoxicity).

43 Leaves completely yellow (MYV).

44 Leaf with one white sector (Chimera).

45 Mottling in chlorotic patches on a melon leaf—Cucumber Mosaic Virus (CMV).

46 Mottling in dark/light green patches on a cucumber leaf—Cucumber Mosaic Virus (CMV).

47 Mottling causing deformation on a courgette leaf—Cucumber Mosaic Virus (CMV).

48 Mottling in small green patches on a courgette leaf—Zucchini Yellow Mosaic Virus (ZYMV).

Mottling can differ in intensity and appearance and is generally more easily observed on young leaves.

After studying these four photographs of leaf mottling you should find it easier to recognise the condition; but you will also realise how difficult it is to associate one type of mottling with one particular virus. This symptom, although a strong indication of virus disease, is not sufficient to allow precise identification of the causal agent. We would advise you to contact a specialist laboratory qualified to carry out the tests necessary for accurate identification of the virus in question.

Mottled leaves

POSSIBLE CAUSES

- **Cucumber Green Mottled Mosaic Virus (CGMMV)** (factfile 26)
- **Cucumber Mosaic Virus (CMV)** (factfile 29)
- **Squash Mosaic Virus (SqMV)** (factfile 27)
- **Zucchini Yellow Mosaic Virus (ZYMV)** (factfile 32)
- **Water Melon Mosaic Virus (WMV2)** (factfile 31)
- **Cucumber Toad Skin Virus (CTSV)** (factfile 34)
- **Papaya Ring Spot Virus (PRSV ex WMV1)** (factfile 30)
- **Viruses causing yellowing** (refer to heading 'Yellow Leaves', page 55)

- *Fusarium oxysporum* **f.sp.** *melonis* (factfile 23)
- *Pseudoperonospora cubensis* (factfile 11)

- **Various phytotoxicities**

It is not always easy to detect mosaic on the leaves of a diseased plant; in full sunlight it is almost impossible. The best way to examine the leaf is with background illumination or in the shade.

ADDITIONAL DIAGNOSTIC GUIDELINES

	Virus transmitted by:		*Fusarium oxysporum* f.sp. *melonis*	Various phytotoxicities
	contact SqMV	aphids CMV, WMV2, ZYMV		
Frequency of disease	Rare	Very frequent	Very frequent	Quite rare
Type of cultivation – open field – protected cover	– (+) +	+ + +	+ + + +	+ +
Spread in field	In rows	A focus or generalised	A focus	One area or general
Stage of symptom appearance	Very early	At any time	Beginning or mid cultivation	At any time
Presence of symptoms on fruit	+ Highly visible	+ Highly visible	– (+)	+ Not obvious
Varietal resistance	–	Tolerant to CMV (cucumber & gherkin)	+	– Susceptibility varies with variety

49 Paler colour between the veins of a melon leaf, with tissues adjacent to veins remaining green (vein banding)—Papaya Ring Spot Virus (PRSV).

50 Courgette leaf with paler colour along the veins—Zucchini Yellow Mosaic Virus (ZYMV).

51 Chlorotic spots at the base of a courgette leaf which is becoming mottled—Cucumber Mosaic Virus (CMV).

52 Dark green bubbles on a melon leaf contrast with the paler green of the rest of leaf giving a 'mosaic' appearance—Zucchini Yellow Mosaic Virus (ZYMV).

Other discolorations of the leaf can appear similar to mosaic; careful examination is required for an accurate diagnosis.

Incidence of viruses

	Very common viruses	Common viruses	Rare viruses
Melon	CMV, WMV2	ZYMV	SqMV, PRSV
Cucumber	CMV	WMV2*, ZYMV	CGMMV, PRSV, CTSV
Courgette	CMV, WMV2* ZYMV		PRSV
Gourd			CMV, WMV2, ZYMV

*Virus not dangerous

Period of spread of virus infections

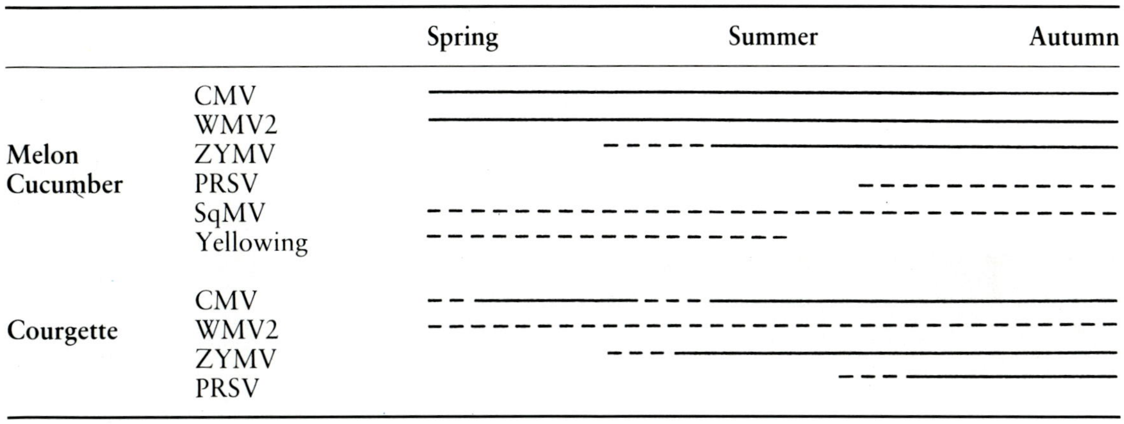

		Spring	Summer	Autumn
Melon Cucumber	CMV			
	WMV2			
	ZYMV			
	PRSV			
	SqMV			
	Yellowing			
Courgette	CMV			
	WMV2			
	ZYMV			
	PRSV			

- - - - Virus rare or not dangerous

——— Virus common

53

53 Mosaic and chlorotic spots, characteristic of the onset of disease—Cucumber Mosaic Virus (CMV).

54

54 Well-developed mosaic, characteristic of a more established disease—Cucumber Mosaic Virus (CMV).

55

55 Diffuse mosaic accompanied by vein banding—Cucumber Mosaic Virus (CMV) + Water Melon Mosaic Virus type 2 (WMV2).

56 Large patches of mottling on apical leaves whose growth has been stunted—Cucumber Mosaic Virus (CMV) + Zucchini Yellow Mosaic Virus (ZYMV).

56

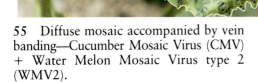

MELON

CMV is the virus that most frequently attacks melon plants, often in association with other less common viruses e.g. Water Melon Mosaic Virus type 2 (55) and Zucchini Yellow Mosaic Virus (56).

MELON

VIRUS	SYMPTOMS	
	on leaves	on fruit
CMV	Mosaic Growth reduction	Marbling. Aborted flowers and reduced yield (**389**)
ZYMV	Prominent veins. Yellowing. Dwarfing, serious deformation, outgrowths (**20–21**) Lethal wilt (**219**) (according to pathotype and variety)	Slight external mottling, external cracks. Internal marbling, hardening of flesh (**381–382**)
WMV2	Mottling with frequent deformation. Vein banding. Reduced leaf area	Pronounced mottling. Fruit sometimes dented (**390**)
PRSV	Generalised yellowing. Deformed leaves with striped appearance.	Mottling with raised dark green patches.
SqMV	Pronounced vein banding followed sometimes by serious distortion. Plant may recover.	Mottling. Reduced yield.

57   Mottled patches, accompanied locally by vein banding—Water Melon Mosaic Virus type 2 (WMV2).

58   Pale green leaf with scattering of darker green 'islets' sometimes angular in shape—Zucchini Yellow Mosaic Virus (ZYMV).

59   Vein banding on slightly wrinkled leaf—Papaya Ring Spot Virus (PRSV).

60   Very marked vein banding also affecting the secondary veins—Squash Mosaic Virus (SqMV).

MELON

Different stages in the development of a virus epidemic transmitted by aphids in a non-persistent manner. Examples of Cucumber Mosaic Virus (CMV) or Water Melon Mosaic Virus type 2 (WMV2) affecting melon plants.

(A)

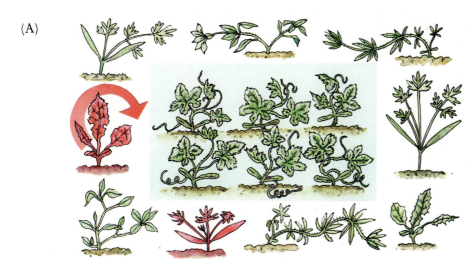

(A) CMV and WMV2 are not transmitted to the melon plant via the seeds. If the plants have been raised in a nursery with protection from aphids, all the plants should be healthy when transplanted. On the other hand, fields are surrounded by weeds some of which can transmit the virus via seeds (e.g. CMV in starwort); others, being frost-resistant, survive the winter and are hosts of the viruses and possibly the aphid vectors.

(B)

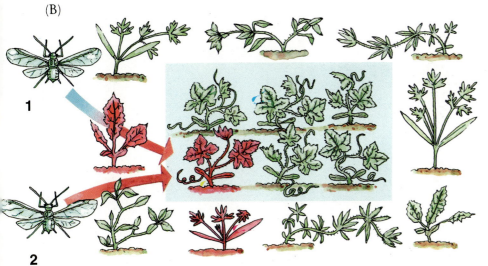

1

2

(B) Infection first occurs a short time after planting. The aphids infest the plot edges: some (1) are not virus vectors when they first appear in the area but they may well become so after alighting on affected weeds, and will then go on to infect the young melon plants. Others (2) have acquired the virus from neighbouring crops and will thus initiate direct infection of the young plants. The first diseased plants in a field will form the primary focus of the disease.

(C)

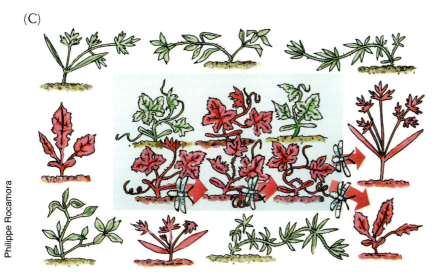

(C) Subsequently, flying aphids will rapidly spread the virus to the whole crop. This is the second stage in the development of an epidemic. It should be noted that a large number of weeds will also be infected; they will form a reservoir of viruses which will survive the winter and contaminate subsequent crops.

Philippe Rocamora

47

61 Very marked mottling on young leaves—Cucumber Mosaic Virus (CMV).

63 Vein banding and mottling on young leaves—Zucchini Yellow Mosaic Virus (ZYMV).

62 Green mottling and distortion in young leaves—Cucumber Green Mottle Mosaic Virus (CGMMV).

64 Progressive yellowing of the veins in an old leaf creating a complete network—Cucumber Toad Skin Virus (CTSV).

CUCUMBER

| VIRUS | SYMPTOMS | |
	on leaves	on fruit
CMV	Mottling with blisters (**13–22**). Reduced growth. Wilting (**218**)	Marked mottling, fruit surfaces slightly dented (**394**)
ZYMV	Prominent veins. Mottling causing greater/lesser deformations (**23**)	Marbling (**395–402**)
WMV2	Marbling. Slight mottling.	No symptoms or very slight marbling.
CGMMV	Mottling [–] green with blistered look on lamina (**24**)	No serious deformation Reduced yield.
CTSV	Prominent veins on old leaves. Contraction and rolling of young leaves, veins become twisted (**15**). Shortening of internodes.	Mottling (**396**) Fruit deformed; corky cracks on surface (**406**).
PRSV	Mottling and deformation of lamina.	Mottling and deformation of fruit.
SqMV	Chlorotic spots. Veins become prominent, then recover.	No symptoms.

COURGETTE

65 Star-shaped yellow mosaic with lamina deformation—Cucumber Mosaic Virus (CMV).

66 Discrete marbling on young leaf becoming blurred as it reaches the adult stage—Water Melon Mosaic Virus type 2 (WMV2).

67 Small, angular mosaic spots scattered over the leaf—Zucchini Yellow Mosaic Virus (ZYMV).

68 Slightly raised, irregular spots or patches of mottling—Papaya Ring Spot Virus (PRSV).

VIRUS	SYMPTOMS	
	on leaves	on fruit
CMV	Mottling, sometimes causing gross deformation. Contracted, rolled leaves close to ground (**14**). Mottling and deformation of petioles.	Mottling causing deformation. Fruit has 'pitted' appearance (**398–405**)
ZYMV	Prominent veins. Mottling, yellowing. Severe foliar deformation (leaves in strips) (**34–35**).	Fruit mottled and dented (**399–405**)
WMV2	Marbling. Lamina serrations are more fretted (this depends on the strain (**32**))	Green spots on courgette with yellow fruit.
PRSV	Mottling. Leaves more/less laciniate with silvering (**26–104**)	Mottled and dented fruit (**405**) Usually no symptoms varieties with green fruit.
SqMV	Mottling may cause deformation, filiform appearance.	Mottling and deformation of fruit which is dented.

69 Very marked vein banding on young courgette leaf—Zucchini Yellow Mosaic Virus (ZYMV).

70 Mottling in small, dull green spots accompanied by vein banding on adult courgette leaf—Zucchini Yellow Mosaic Virus (ZYMV).

71 Discrete mottled spots on water melon leaf—Zucchini Yellow Mosaic Virus (ZYMV).

72 Young squash leaf mottled and partially deformed—Zucchini Yellow Mosaic Virus (ZYMV).

GOURD, WATER MELON, SQUASH

Other diseases cause symptoms which, as they develop, could be mistaken for mosaic and mottling. This is the case, for example, with the two diseases caused by fungi which are described below. Care must be taken to avoid confusion.

● *Fusarium oxysporum* f.sp. *melonis*

On plants affected by the yellowing forms of *Fusarium oxysporum* f.sp. *melonis* (a vascular wilt fungus) it is possible to see a fairly diffuse and heterogeneous yellowing of the laminae of the first leaves to show symptoms (73). This can be mistaken for mottling or mosaic caused by a virus. A more thorough examination must be done to confirm the presence or absence of this pathogen, particularly in the vascular tissue of the stem (see page 146).

● *Pseudoperonospora cubensis*

When conditions are very favourable for its development downy mildew, which affect Cucurbitaceae, cause numerous spots, at first oily then yellow, on the leaves (74). These are often angular, particularly in cucumbers. The lamina appears to be covered with mosaic spots which must not be confused with those caused by a virus (see pages 98–100).

73

73 Uneven yellowing of lamina giving a mottled appearance to this melon leaf—*Fusarium oxysporum* f.sp. *melonis*.

74

74 Mosaic in well defined patches associated with the veins on a cucumber leaf—*Pseudoperonospora cubensis*.

53

75 Inter-vein chlorosis on leaves which are becoming thicker and brittle—Luteovirus of Melon.

76 Yellowing in almost all leaves; only the youngest remain green—Melon Yellows.

77 Different aspects of yellowing: diffuse in young leaves, generalised in older leaves—Melon Yellows.

78 First signs of yellowing; symptoms become more marked with time—Lettuce Infectious Yellow Virus (LIYV).

MELON

These symptoms are generally linked to very heavy infestations of the vectors (aphids, whitefly) on the preceding crop, on the growing crop or on nearby crops.

Leaves partially or totally yellow or chlorotic

POSSIBLE CAUSES

- **Parasitic fungi and bacteria in the roots, collar, stem and vessels of the Cucurbitaceae** (see relevant sections, in particular *Fusarium oxysporum* **f.sp.** *melonis*)
- **Luteovirus of melon** (factfile 33)
- **Cucumber Yellow Virus (CYV)** (factfile 33)
- **Musk Melon Yellow Virus (MYV)** (factfile 33)
- **Lettuce Infectious Yellow Virus (LIYV)** (factfile 33)
- **Cucumber Mosaic Virus (CMV)** (factfile 29)
- **Zucchini Yellow Mosaic Virus (ZYMV)** (factfile 32)
- **Cucumber Toad Skin Virus (CTSV)** (factfile 34)

- **Nutritional deficiencies**
- **Excess salinity**
- **Hypersusceptibility to mildew**
- **Physiological yellowing of courgette leaves**
- **Various phytotoxicities**

See also the section 'Wilted, desiccated leaves (preceded or accompanied by yellowing)'.

Yellowing or chlorosis of leaves is a symptom often found by growers of cucurbitatious crops and may appear in many forms:

- Restricted to a small area in the form of a spot (see page 101) or in association with a spot, surrounding it with a more or less marked halo (e.g. page 74);
- Affecting one side only of a leaf; this unilateral yellowing is often characteristic of vascular diseases (see page 148)

- Developing from the veins outwards, or between the veins (inter-vein yellowing).

It may first become apparent on the young apical leaves or on the old leaves at the plant base; in certain cases the intermediate leaves may be affected. Sometimes the condition affects the entire plant. Intensity of the colour varies, and may cause actual blanching of the leaves.

It is a 'key symptom' of a malfunction in the plant, and often follows:
- Non-parasitic diseases such as deficiencies or phytotoxicities;
- One or several pathogens causing diseases which are localised either on the leaves or in other parts of the plants, particularly roots and stems.

It is rarely easy to determine the cause or causes: study the situation very carefully before making an assumption.

ADDITIONAL DIAGNOSTIC GUIDELINES

- Luteovirus of melon
- Musk Melon Yellow Virus (MYV)
- Lettuce Infectious Yellow Virus (LIYV) and other viruses

As you can see from the photographs opposite, the symptoms caused by these three viruses are very similar. In addition, each virus has a slightly different effect depending on the plant variety. It is therefore extremely difficult to establish an accurate diagnosis; it is often possible only to make an assumption. Refer to the table on page 57 to discover whether or not your assumption is correct.

79 Early yellowing can be confused at this stage with a mosaic disease; the very young leaves are symptom-free—Cucumber Yellow Virus (CYV).

80 Yellowing is becoming more intense and the presence of spots retaining their deep green colour is characteristic of the disease—Cucumber Yellow Virus (CYV).

81 · The leaf may become almost completely yellow, it wilts, and the lamina becomes brittle as it thickens—Cucumber Yellow Virus (CYV).

82 Yellowing always affects old leaves more severely—Cucumber Yellow Virus (CYV).

CUCUMBER

In the past it was often thought that Melon and Cucumber Yellows were due to nutrient deficiencies, but today in an increasing number of cases viruses are found to be the cause.

In France, yellowing has recently been observed in field cultivated crops, or in plant-breeding glasshouses in summer or autumn. A virus from the **Luteovirus** group, transmitted by aphids in a persistent manner (the first example reported affecting Cucurbitaceae), was found to be associated with this disease (75). In the USA some years ago a disease with symptoms similar to musk melon yellow caused serious damage to melon crops in Southern California. The cause of the disease was Lettuce Infectious Yellow Virus (LIYV) (78), which is transmitted by the whitefly *Bemisia tabaci*. This virus is also thought to be present in the Middle East and there is a danger that *it will reach Europe* if *Bemisia tabaci* infestation reaches serious proportions.

Finally, it should be remembered that other viruses can also cause, among other symptoms, the yellowing of old leaves: for example, Cucumber Yellow Vein Virus (CYVV) is very prevalent in the Middle East and is transmitted by *Bemisia tabaci* while Zucchini Yellow Mosaic Virus (ZYMV) is transmitted by aphids.

Plant part affected	Viruses producing yellowing	Mosaic type viruses
Apex, young leaves	−/+	+++
Old leaves	+++	+/++
Fruit	−/+	+/+++

	Melon luteovirus	Cucumber or Melon Yellow Virus	Lettuce Infectious Yellow Virus
Vectors	Aphids	Whitefly (*Trialeurodes vaporariorum*)	Whitefly (*Bemisia tabaci*)
Susceptible hosts	melon, cucumber courgette	melon, cucumber	melon, cucumber
Type of crops affected	mostly field crops	under cover	field crops

Where crops are cultivated under cover and show yellowing symptoms, heavy infestations of vectors (aphids, whitefly) can strongly suggest a diagnosis of the yellowing diseases.

83

83 Arrested growth and yellowing of apical leaves in a melon plant—Zucchini Yellow Mosaic Virus (ZYMV).

84

84 Yellowing and shrivelling of recently affected young courgette leaves—Cucumber Mosaic Virus (CMV).

85

85 Yellowing in some sectors of cucumber leaves is likely to spread later over the whole surface of the laminae—Cucumber Toad Skin Virus (CTSV).

Other viruses may also cause leaf yellowing, particularly Zucchini Yellow Mosaic Virus (83), Cucumber Mosaic Virus (84) and Cucumber Toad Skin Virus (85). In general, yellowing is first observed on the young leaves. Where Cucumber Toad Skin Virus is concerned, it affects the older leaves first while the young leaves show a characteristic prominence of the veins (64).

To make sure your diagnosis is correct, refer to pages 45, 49 and 51.

- **Nutritional deficiencies**

On the question of deficiencies, there is often a tendency to associate true deficiencies with induced deficiencies.

True deficiencies (where the soil lacks certain elements) are now seldom found, especially with regard to trace elements. Their diagnosis by sight alone is very difficult, because without exception the symptoms they cause are discolorations, deeper or paler yellowing of leaves, which are very difficult for a non-specialist to judge. As a guide, refer to pages 60 and 61 which describe the main symptoms of deficiencies affecting melons and cucumbers.

We more often have to deal with induced deficiencies (where the elements are present but not available), which makes diagnosis no easier. In addition to identifying which element is deficient and sorting out the confusion of possible symptoms, the cause or causes of the deficiency must be determined. There can be various causes: for example, mismanaged irrigation (under-watering or over-watering), soil temperature or pH too high or too low, roots in a poor state, etc.

When you are confronted with symptoms such as these, do not immediately jump to the conclusion that they are due to a deficiency **before you have consulted a specialist and carried out the essential physico-chemical analyses of the soil, the plants, etc.**

Deficiencies occur most commonly in crops whose management has been based on experience only, and where analysis of the soil or nutrient solutions has not been conducted.

Conditions under which the principal induced deficiencies can occur are as follows:

- Retarded development or poor condition of the root system;
- Soil temperature too low;
- Night temperature too high;
- Too much, or badly balanced, fertiliser, often the cause of interaction between elements;
- Inclusion of fermentable organic material;
- Insufficient light;
- Incorrect irrigation (excess water causing asphyxia, lack of water);
- Heavy leaching;
- Soil too chalky or too acid;
- Excess of heavy metals in soil.

The photographs on these two pages illustrate melon and cucumber plants suffering from a 'nutrient problem'. Generally, when a plant shows symptoms of nutrient deficiency, the symptom or symptoms appear first on the youngest or the oldest leaves. To an inexperienced observer they look very similar, making identification almost impossible.

Symptoms of the principal deficiencies occurring in melon and cucumber plants.

	Melon	Cucumber
Nitrogen	Diffuse yellowing of the lamina and the veins in old leaves. Fruit reduced in size (elongated, paler, flesh is pale and tasteless)	Retarded growth. Gradual yellowing of the entire lamina in old leaves. Slimmer fruit, pinched at stylar end, blossom drop.
Phosphorus	Shortening of internodes, dwarfing of plants. Base leaves are reddish-green brown inter-veined necrosis with yellow halo. Smaller fruit with reddish flesh.	Plants do not grow strongly. If deficiency is very severe: young leaves are smaller, dull green. Oily spots becoming necrotic, brown on old leaves.
Potassium	Brown discoloration on edges of young leaves (parasol appearance). Fruit has gritty, bitter flesh.	Shortening of the internodes, young leaves are small, lustreless. Yellowing occurs first around the periphery of the lamina then of inter-vein tissues. Veins remain green a long time. Irregular-shaped fruit, clubbed and pointed.
Iron	Inter-veinal chlorosis of the lamina of young leaves, less accentuated along the veins.	Sometimes growth stops. Inter-veinal yellowing of the young leaves which turn lemon-yellow to yellowish white and become necrotic.
Magnesium	Yellowing of basal leaves starting at the periphery of the lamina, occurring after pricking-out and/or during growth, sometimes leading to serious defoliation.	Inter-vein chlorosis (sometimes in large spots) starting on the periphery of the lamina, resulting in its necrosis. Large veins remain green, leaves are thicker and brittle.

88
89

Do not attempt to solve your problem by relying on past experience alone. Have the necessary analyses carried out.

	Melon	Cucumber
Molybdenum	Retarded growth. Leaves pale green then ivory-yellow, gradually becoming desiccated around the periphery of lamina. Particularly serious in the 'red soils' of the Mediterranean countries and found typically on cauliflowers	Initial slight yellowing of lamina, especially the inter-vein tissues; the condition deteriorates and the leaves die. Flowers are smaller.
Boron	Lower quality fruit.	Growth stops, plants are dwarfed. Young leaves become rolled and desiccated. Old leaves may curve inward and be slightly marbled.
Calcium	Arrested growth of terminal bud which becomes desiccated. Increased number of glassy looking fruit.	Retarded growth, short internodes. Small young leaves have a tendency to curve inwards and become yellow, with small white spots. Old leaves may be rolled. Fruit small, wrinkled and tasteless.
Sulphur		Retarded growth Young leaves paler, small, with tendency to shrivel, edge of lamina more serrated.
Manganese		Yellowing on inter-veinal tissue on lamina which turns gradually yellowish-white with some local necrosis.
Zinc		Short internodes and stunted plants. Inter-veinal marbling of old leaves.
Copper		Retarded growth. Young leaves small. Old leaves lustreless with inter-veinal chlorotic spots followed by necrosis.

90   Yellowing starting on the leaf veins.

91   Diffuse, scattered yellowing of a leaf.

92   Homogeneous yellowing of part of a leaf.

93   Yellowing around the periphery of a leaf lamina; copper poisoning may be the cause of this symptom.

Symptoms of phytotoxicity

MELON

- **Various phytotoxicities**

We suggest that you study some photographs of melon plants which illustrate the type and the spread of the yellowing caused by phytotoxicities. Details of other symptoms arising from these conditions are given on pages 34 and 35. Symptoms which affect other members of the Cucurbitaceae, particularly the cucumber, are often similar to those appearing on melon plants.

Further diagnostic guidelines can be found on page 35.

In addition to photographs, we have listed below the symptoms to look for, as they are described in the literature or observed on cucumber or melon plants accidentally contaminated by products from which the main types of agricultural herbicides are constituted:

- Total cessation or serious retardation of growth making the plant appear dwarfed; the internodes are often very short;
- Proliferation of the axillary branches;
- Leaf roll, more or less marked;
- Incurved, spoon-shaped leaves;
- Thickening and curvature of the stem which may sometimes be blistered;
- Edge of lamina slightly more serrated;
- Diffuse yellowing of entire lamina similar to a mosaic disease;
- Yellowing in more or less well-defined spots;
- Yellowing of the veins, spreading to entire lamina with subsequent necrosis;
- Rapid yellowing and blanching of young apical leaves;
- Blanching of the veins on the periphery of the lamina;
- Inter-veinal yellowing in dirty-grey to bronze patches;
- Rapid inter-veinal yellowing and desiccation;
- Very rapid wilting and desiccation of plants;
- Necrosis and desiccation round the periphery of the lamina of young or old leaves;
- Appearance of small white necrotic spots between the veins;
- Wilting and desiccation of the apex;
- Inter-veinal desiccation of old leaves.

94 Necrosis of the periphery of the lamina of a melon leaf, accompanied by yellowing—Phytotoxicity (Excess salinity).

95 Progressive yellowing and necrosis of the periphery of the lamina of a cucumber leaf—Phytotoxicity.

96 A diffuse yellow patch on the lamina of a courgette leaf—Physiological yellowing.

97 Yellowing of lamina and veins, accompanied by necrosis, in a cucumber leaf—Hypersusceptibility to mildew.

(If necrosis is observed on the periphery of the lamina, refer also to page 111.)

● **Excess salinity**

Cucurbitaceae are moderately susceptible to salinity. When salinity reaches too high a level (e.g. excessive applications of fertiliser), the roots of the plants become slightly 'burnt', with the result that plant growth may be retarded or checked. In addition, the periphery of the laminae of the leaves may become yellow and necrosed (**94**). The latter symptom can also occur where too much manure has been added to the planting trench.

● **Other phytotoxicities**

The addition of fungicides to the root zone is now common practice, for both soil-less and soil cultivation. Often the compounds used have not been specifically formulated for this purpose. The application rates may be too high or there may be an accumulation of residual chemicals after several successive applications.

    The resulting phytotoxicity frequently causes chlorosis round the edge of the lamina, particularly when benomyl and propamocarb hydrochloride have been used (**95**). (Excess boron produces similar symptoms of phytotoxicity.)

● **Physiological yellowing in courgettes**

In certain conditions, which have not yet been properly defined, old courgette leaves show diffuse yellow patches often starting at the base of the main veins (**96**). This discoloration is not caused by a pathogen; it appears to be irreversible but harmless to growth and plant yield.

● **'Hypersensitivity' to mildew**

There are now some varieties of melon and cucumber which are tolerant to the powdery mildews affecting the cucurbits (*Sphaerotheca fuliginea* and *Erysiphe cichoracearum*—**193**, **194**, **195**, page 102). In specific environmental conditions, there may be a reaction resulting in the appearance of numerous small necrotic flecks on the leaves (**97**), indicating a hypersensitivity response by the plant tissues (see Appendix III).

● **Various fungal diseases**

Several fungal diseases cause total or partial yellowing of the leaves in affected plants. This is particularly the case with the vascular wilt fusarium diseases of cucurbits (**98**). In addition to yellowing, there are very often other, more characteristic symptoms (on the roots, in the vessels and on the stem, etc.) which help to distinguish the diseases dealt with in this section.

98

98 Yellowing, sometimes in sectors, of the lamina and veins of melon leaves—*Fusarium oxysporum* f.sp. *melonis*.

99 Rapid yellowing and blanching affecting particularly the young leaves of melon—Phytotoxicity.

100 Small pale green to white patch on melon leaf—Chimera.

101 Cucumber leaf with clearly defined pale yellow patch on lamina—Chimera.

102 Large white sectors on cucumber leaf—Chimera.

103 Diffused bleaching affecting much of the lamina of the water melon leaf—Chimera.

Leaves partially or totally bleached

POSSIBLE CAUSES

- Sectorial chimera
- Various phytotoxicities

ADDITIONAL DIAGNOSTIC GUIDELINES

- **Various phytotoxicities**

Certain herbicides disturb the plants' photosynthetic activity and are responsible for yellowing of the leaves, followed very rapidly by bleaching (**99**). The same symptoms can result from the application of some mixtures of pesticides. They may also be accompanied by further symptoms which are described on pages 34, 62, 63, etc. Conditions favourable to their development are noted on page 35.

- **Sectorial chimera**

Growers may sometimes observe bleached or yellow sectors, more or less well defined, on the laminae of one or several leaves of some of the plants (**100, 101, 102, 103**). This discoloration can also affect the stem and the fruit (see **397**). The cause is a localised genetic change in the leaf tissues which also impairs the development of the leaf or leaves which in turn may become deformed. This non-parasitic problem is found in all cultivated varieties of cucurbits especially at low temperatures; certain varieties are particularly at risk.

104 Courgette leaf almost totally 'silvered' —Papaya Ring Spot Virus.

105 'Silvered', angular patches, more/less defined by the veins, on a courgette leaf— Physiological blotches.

106 Paling of the veins and inter-vein tissues on a melon leaf, which is acquiring a 'bronzed' appearance—Phytotoxicity.

107 Large number of small necrotic spots, giving a 'bronzed' appearance to this melon leaf—Phytotoxicity.

Leaves 'silvered', 'bronzed'

POSSIBLE CAUSES

- **Papaya Ring Spot Virus (PRSV ex WMV1)** (factfile 30)
- **Spider mites** (see Appendix II)
- **Thrips** (see Appendix II)

- **'Silvering' of courgettes**
- **Physiological blotches of courgettes**
- **Various phytotoxicities**

ADDITIONAL DIAGNOSTIC GUIDELINES

- **Papaya Ring Spot Virus**

In addition to the symptoms of mottling (**68**) and deformation of the lamina (**33**), the leaves of courgette plants affected by this virus will often take on a very characteristic 'silvered' appearance (**104**) which aids diagnosis.

- **Physiological blotches**

In certain varieties of the courgette and gourd, angular, 'silvered' patches on the leaves occur very frequently (**105**). They are caused by localised detachment of the epidermis and the 'silvered' look is due to a layer of air between the epidermis and the parenchyma. Their frequency varies considerably according to the variety of courgette or gourd and the climatic conditions.

- **Various phytotoxicities**

Some pesticides (herbicides absorbed by the roots (**106**), fungicides or insecticides applied too generously or in unsuitable conditions (**107**)) cause 'bronzing' of the leaves. This is often accompanied by other symptoms; these, and the conditions affecting their development, are described on pages 34, 35, 62, 63, etc.

- **'Silvering' in courgettes**

This new disease (not to be confused with physiological blotches) has been observed for some years in southern USA (Florida) and in the Middle East. The first sign is that the veins become prominent; subsequently the plant appears to be 'silvered' all over but without distortion. The leaves are uniformly 'silvered' on their upper surface, although the underside retains its normal colour. The fruit may also be paler in colour and yield is likely to be lower. Recently the disease has been reported in the south-east of France (Berre region).

It is apparently caused by the nymphs of the whitefly *Bemisia tabaci* feeding on plants. The saliva of the nymphs contains a 'toxic compound' which induces the symptoms. The proof that a virus is responsible for the disease has not been obtained. If a virus is involved with 'silvering' particularly when excessive numbers of *Bemisia* nymphs are feeding on the plants, an effective insecticide should be able to control the disease.

Spots on leaves

SYMPTOMS EXAMINED

- Smallish spots, with or without necrosis, on leaves
- Spots with brown areas often at edge of leaves, sometimes with brown curving patterns
- Oily spots on leaves (subsequent yellowing and necrosis)
- Yellow spots on leaves
- Powdery white spots on leaves

POSSIBLE CAUSES

- *Pseudomonas syringae* pv. *lachrymans*
- *Pseudomonas syringae* (group II)
- *Xanthomonas campestris* pv. *cucurbitae*

- *Acremonium* sp.
- *Alternaria alternata (Alternaria tenuis)*
- *Alternaria cucumerina*
- *Alternaria pluriseptata*
- *Ampelomyces quisqualis*
- *Botrytis cinerea*
- *Cercospora citrullina (Cercospora cucurbitae)*
- *Cladosporium cucumerinum*
- *Cladosporium* sp.
- *Colletotrichum lagenarium*
- *Corynespora cassiicola (Cercospora melonis)*
- *Didymella bryoniae*
- *Erysiphe cichoracearum*
- *Helminthosporium cucumerinum*
- *Leandria momordica*
- *Leveillula taurica*
- *Myrothecium roridum*
- *Penicillium* spp.
- *Phytophthora capsici*
- *Pseudoperonospora cubensis*
- *Septoria cucurbitacearum*
- *Sphaerotheca fuliginea*
- *Stephanoascus* sp.
- *Tilletiopsis* spp.
- *Ulocladium atrum (Ulocladium cucurbitae)*
- *Verticillium lecanii*

- Melon Necrotic Spot Virus (MNSV)
- Tobacco Necrosis Virus (TNV)
- Other necrotic viruses

- Adverse climatic conditions
- Intumescence
- Various phytotoxicities
- Various physiological reactions
- Pollen deposits

SIMPLE DIAGNOSIS

Diseases responsible for leaf spots are often quite easy to identify, providing that you follow these recommendations:

- Collect and examine several leaves from different plants before making a diagnosis;
- Always look at the underside of the leaves and note the presence of any fungal fructifications or other factors which would help to confirm the diagnosis.

OBSERVATION GUIDE

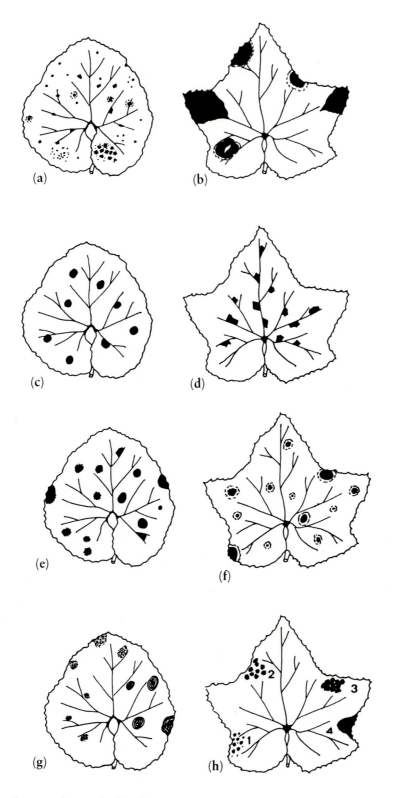

Size, appearance, shape and spread of leaf spots

(a) Pinpoint or small spots;
(b) Localised patches round the edge of the lamina;
(c) Round spots;
(d) Angular spots bounded by the veins;
(e) Diffuse spots (on left) or well-defined spots (on right);
(f) Spots with haloes;
(g) Powdery spots (on left) or in rings or concentric circles (on right);
(h) Small group of spots (1), enlarging (2, 3) and fusing together (4).

72

Examples of leaf spots

108 Small, brown necrotic spots on a melon leaf—*Cladosporium cucumerinum*.

109 Brown patches on the periphery of lamina, with yellow haloes, on a cucumber leaf—*Didymella bryoniae*.

110 Angular spots, oily, turning yellow, on a cucumber leaf—*Pseudoperonospora cubensis*.

111 Powdery patches on a melon leaf—*Sphaerotheca fuliginea*.

112 Greyish spots with brown periphery, sometimes angular, with yellow halo—*Cladosporium cucumerinum*.

114 Brown, necrotic spots sometimes stretching along the veins—*Colletotrichum lagenarium*.

113 Necrotic brown veins on leaf; an oily exudate can be seen on the developing oily and elliptical canker—*Cladosporium cucumerinum*.

115 Numerous oily spots, more or less elongated, and covered with a brown exudate —*Colletotrichum lagenarium*.

116 Fawn spots, confluent in places, whose tissues are becoming desiccated and are splitting—*Didymella bryoniae*.

MELON

Smallish spots, with or without necrosis, on leaves

POSSIBLE CAUSES

- *Pseudomonas syringae* pv. *lachrymans* (factfile 1)
- *Pseudomonas syringae* (group II) (factfile 4)
- *Xanthomonas campestris* pv. *cucurbitae* (factfile 2)

- *Alternaria alternata (Alternaria tenuis)*
- *Alternaria cucumerina* (factfile 12)
- *Alternaria pluriseptata*
- *Cercospora citrullina* (factfile 12)
- *Cladosprium cucumerinum* (factfile 5)
- *Colletotrichum lagenarium* (factfile 6)
- *Corynespora cassiicola* (factfile 12)
- *Helminthosporium cucumerinum*
- *Leandria momordica*
- *Myrothecium roridum*
- *Phytophthora capsici* (factfile 14)
- *Septoria cucurbitacearum* (factfile 12)
- *Ulocladium atrum (Ulocladium cucurbitae)* (factfile 12)

- **Melon Necrotic Spot Virus (MNSV)** (factfile 28)
- **Tobacco Necrosis Virus (TNV)** (factfile 28)
- **Other necrotic viruses** (factfile 28)
- **Various phytotoxicities**

ADDITIONAL DIAGNOSTIC GUIDELINES

	Cladosporium cucumerinum	*Colletotrichum lagenarium*	Virus trans- mitted by *Olpidium* spp.	Phytotoxicity
Hosts attacked	melon courgette	melon	melon cucumber	melon cucumber courgette
Frequency of outbreaks				
field	++	+	−/+	+
under cover	+	−	+ serious in soilless culture	+
Distribution of spots on leaves and on plant	+/− localised	+/− localised	+/− generalised	see p.14
Presence of symptoms				
on stems	+	+	+/−	+/−
on fruit	+	+	+/−	+/−

In melon plants, the first signs of disease caused by *Didymella bryoniae* are more or less circular spots, fawn to black, 5 mm in diameter, sometimes with a yellow halo (**116**). Later, these spots become desiccated and split. These spots are rarely seen in French melon crops.

117

117 Small, depressed, beige, 'lip'-shaped canker, its centre covered with dark green—*Cladosporium cucumerinum.*

118

118 More/less corky patches, elongated; fungal fructification makes the centres of the patches appear dark green to black—*Cladosporium cucumerinum.*

119

120

119 Elongated, confluent impaired tissues, pink > orange, covered with numerous pinkish pinpoints (acervuli)—*Colletotrichum lagenarium.*

120 Oval patches, depressed, covered with black pits and salmon-pink gelatinous masses (acervuli)—*Colletotrichum lagenarium.*

MELON

121 Numerous spreading brown patches, with beige centres—*Colletotrichum lagenarium*.

122 Later, these patches become desiccated and split—*Colletotrichum lagenarium*.

123 Numerous small, very elongated cankers along the stem—*Colletotrichum lagenarium*.

124 Developing young fruit showing curvature caused by a discrete beige canker—*Colletotrichum lagenarium*.

77

125 Spots with yellow haloes; portions of the grey, necrotic tissues at the centres have dropped—*Cladosporium cucumerinum.*

126 Detail of a spot on the leaf—*Cladosporium cucumerinum.*

127 Numerous greasy, necrotic spots on the petiole—*Cladosporium cucumerinum.*

128 Small greasy spots of canker on fruit—*Cladosporium cucumerinum.*

COURGETTE

129 Brown and necrotic patches on a young leaf—*Phytophthora capsici.*

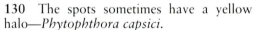

130 The spots sometimes have a yellow halo—*Phytophthora capsici.*

131

131 The apical leaves and the very young fruits are particularly susceptible; they rapidly become brown and necrotic—*Phytophthora capsici.*

132 The petals are also susceptible and become brown; the young fruit may show sticky exudates—*Phytophthora capsici.*

COURGETTE

133 Brown to black spots on leaves are discrete at first, becoming confluent in places—*Cladosporium cucumerinum*.

134 Early, 'greasy', elliptical cankers on the stem, covered with green mould—*Cladosporium cucumerinum*.

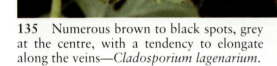

135 Numerous brown to black spots, grey at the centre, with a tendency to elongate along the veins—*Cladosporium lagenarium*.

136 Elongated brown impaired tissues, confluent in places, covered with salmon pink pitting (acervuli)—*Cladosporium lagenarium*.

137 Oily concave patches, covered with small pink gelatinous masses (acervuli)—*Cladosporium lagenarium*.

138 Leaves covered with small necrotic lesions; note the presence of a wilting young leaf—Melon Necrotic Spot Virus (MNSV).

139 In the early stages these necrotic spots are oily; later a greasy halo persists, visible on the underside of the leaves—Melon Necrotic Spot Virus (MNSV).

140 Later, the lesions spread giving the necrotic spots a rusty colour—Melon Necrotic Spot Virus (MNSV).

141 Sometimes the lesions spread along the veins—Melon Necrotic Spot Virus (MNSV).

MELON

Several of the viruses transmitted by soil fungi belonging to the genus *Olpidium* can be responsible for necrotic spots on leaves; these can easily be confused with the spots caused by certain parasitic fungi. **Examine the evidence carefully.**

At present, there are at least six viruses which may be responsible for causing necrosis of melon and cucumber leaves. Their symptomatology is similar, making a visual diagnosis almost impossible. To add to the difficulty, there may be wide variations in the symptoms caused by the same virus according to the season (damage is generally more severe in winter or spring) and the variety. *In Europe*, two viruses are common: **Melon Necrotic Spot Virus** (138, 139, 140, 141, 142) and **Tobacco Necrosis Virus** (143, 144, 145, 146, 147).

CUCUMBER

142

142 Necrotic spots, darker brown on the periphery, becoming larger and more confluent in places to form patches of dead tissue—Melon Necrotic Spot Virus (MNSV).

143

143 Frequently, entire sectors of the lamina and whole leaves become necrotic and die—Tobacco Necrosis Virus (TNV).

144 Necrosis is sometimes localised in the veins, and in particular on the periphery of the lamina—Tobacco Necrosis Virus (TNV).

145 Small longitudinal splits are visible on the petioles and stalks—Tobacco Necrosis Virus (TNV).

146 These splits are less frequently seen along the leaf veins—Tobacco Necrosis Virus (TNV).

147 Necrosis developing on a stub left after pruning; the virus has been transmitted from plant to plant during thinning-out—Tobacco Necrosis Virus (TNV).

CUCUMBER

148 Cucumber leaves with grey, dry, angular spots, oily at the start of the outbreak—*Pseudomonas syringae* pv. *lachrymans*.

150 Greyish spots on cucumber leaves; the desiccated tissue is splitting and falling out—*Pseudomonas syringae* pv. *lachrymans*.

151 Brown necrotic patches, sometimes following the veins, on a melon leaf—*Pseudomonas syringae*.

149 Detail showing recently developed moist, angular spots—*Pseudomonas syringae* pv. *lachrymans*.

152 Brown necrotic lesion on a melon stem, located near the tie where water becomes trapped—*Pseudomonas syringae*.

Bacterial diseases of the cucurbitaceae

• *Pseudomonas syringae*
This differs from the preceding *Pseudomonas* in the symptoms it causes in melons (151, 152). Outbreaks sometimes occur in crops grown under cover, particularly in the Almeria region of Spain. The high humidity conditions in this production area, close to the sea, are decidedly favourable to its development. As the organism is very similar to *Pseudomonas syringae* pv. *lachrymans*, it should be possible to control it using the same methods (see factfile 1).

• *Xanthomonas campestris* pv. *cucurbitae*
It is very easy to overlook the spots on courgette leaves caused by this organism (153, 154) since although they are similar to those caused by *Pseudomonas syringae* pv. *lachrymans*, they are smaller. It is the symptoms affecting the fruit (323, 324, 325, 326) that growers should be looking out for as these cause serious problems of rotting during storage (see page 162). It seems that according to the conditions, the symptoms may appear on the fruit or the leaves, but rarely on both.

This bacterium is transmitted via seeds and there is little known about its biology. The same control measures that are recommended for *Pseudomonas syringae* pv. *lachrymans* can be used.

153

153   Small, angular, yellowish to grey spots (greasy at the start of the outbreak), on a gourd leaf—*Xanthomonas campestris* pv. *cucurbitae*.

154

154   Detail showing spots with a tendency to become confluent and necrotic, particularly at the leaf edge in places where water accumulates—*Xanthomonas campestris* pv. *cucurbitae*.

155 Small, slightly angular spots, brown on the periphery, with a yellow halo, on a cucumber leaf—*Ulocladium atrum*.

156 These spots, greasy at the start of the outbreaks, may spread if water lies on the leaf; necrosis will follow—*Ulocladium atrum*.

157 Small, elongated spots, beige/reddish-brown, on a cucumber stem—*Corynespora cassiicola*.

158 Numerous spots, circular to angular, brown at first, then reddish-brown to beige, on cucumber leaves—*Corynespora cassiicola*.

CUCUMBER

- *Ulocladium atrum/Ulocladium cucurbitae*

This fungus is rarely seen in crops grown in France. The fungus was isolated several times during the years 1985–86 in eastern France (Luneville region) (155, 156) from cucumber crops grown under cover. This is the only case known so far. In England, where the fungus has been reported for the first time, it does not appear to cause much damage to cucumber crops. The disease occurs rarely and epidemics are not widespread (1). When melon plants were artificially inoculated, the organism was found to be pathogenic; but it has been found occurring naturally on melons.

- *Corynespora cassiicola (Cercospora melonis)*

Corynespora cassiicola (also known as *Cercospora melonis*), like the fungus mentioned above, is quite rare in northern Europe. Although it has a wide host range, it attacks mainly cucumbers (the few varieties which are still susceptible) (Var) (157, 158). It does not affect the cucumber fruits although fruits of related genes are affected.

- *Alternaria cucumerina*

There has been no serious outbreak of this fungus in France during the last decade; it now appears to have completely vanished. It can cause leaf spots which at first are small and round and then develop rapidly creating an appearance of concentric circles on the upper side. The melon and water melon are more susceptible; in water melons the spots are dark brown to black, whereas on the melon and cucumber they are paler. The fungus is also reported to cause deterioration in over-ripe field-cultivated fruit and also of fruit during storage. Another *Alternaria (Alternaria pluriseptata)* has been found on cucumbers in Czechoslovakia.

- *Septoria cucurbitacearum*

This is another parasitic fungus which is no longer found in France. It also causes small round spots on the upper surface of the leaf which are olive brown, slightly raised and 3–5 mm in diameter (can reach up to 8 mm). On old spots, paler and desiccated, black dots appear which are fungal fructifications (pycnidia).

- *Cercospora citrullina (Cercospora cucurbitae)*

Entrenched in tropical and sub-tropical regions, this fungus causes leaf spots on many species of the Cucurbitaceae. On the cucumber, melon and courgette the spots can reach 0.5–10 mm in diameter. They are greyish ochre, becoming transparent. Leaf fall may occur. On water melons, the spots affect the young leaves first and are generally smaller.

- *Alternaria alternata (Alternaria tenuis)*

In Crete this organism mainly affects cucumber crops grown under cover. The spots are at first round and yellow, then increase up to about 5 cm in diameter. Within a short time they become light brown, necrotic and papery, with a yellow margin. Brown/black fungal fructifications appear on the necrotic lesions. Fungicides used for the treatment of *Alternaria cucumerina* are equally effective on this organism.

(1) On the other hand, in the USA, a probably identical *Ulocladium* known as *Ulocladium cucurbitae* can cause considerable damage to late season young and established plants. The *Ulocladium* observed in France is morphologically comparable to the two species recognised in England and the USA; we think it is probably *Ulocladium consortiale*.

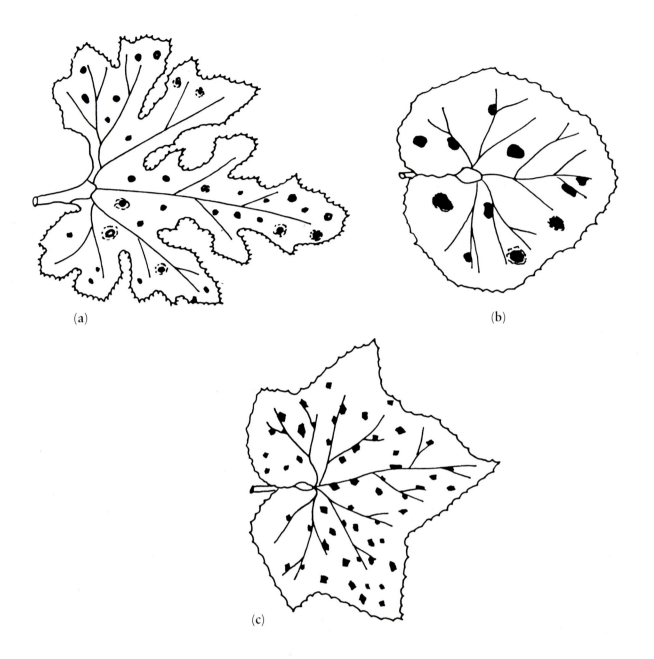

(a) Spots caused by *Cercospora citrullina* on a courgette leaf.

(b) Spots caused by *Alternaria cucumerina* on a melon leaf.

(c) Spots caused by *Leandria momordica* on a cucumber leaf.

- *Myrothecium roridum*

This fungus attacks melons in southern Texas (USA). Its main target is the fruit (see page 187) but it is also the cause of round, irregular spots on the leaves, 2–15 mm in diameter. The centre is often amber coloured, surrounded by a brown margin in the form of concentric rings, which may lead to their being mistaken for the spots caused by *Alternaria cucumerina*. There are also numerous dark green cushions, the sporodochia of the fungus. Hollow, necrotic, elongated lesions may occur on the stems and petioles.

- *Leandria momordica*

Formerly known as *Stemphylium cucurbitacearum* this fungus produces angular spots very similar to those caused by *Corynespora cassiicola*. In time the lesions become circular and the veins become dark red. Only a few countries have reported the presence of this organism: in Brazil it has affected *Momordica charantia* and in the USA (in Florida) it has affected cucumbers, although the damage is rarely serious.

- *Helminthosporium cucumerinum*

In Japan, outbreaks of this fungus seem to affect mainly cucumber plants. The resulting small spots never cause much damage to the crops.

159 Light reddish-brown spots, browner on the periphery—Phytotoxicity.

160 Numerous beige, confluent, elongated chlorotic spots, localised around the leaf veins—Phytotoxicity.

161 Beige/yellowish patch, more or less well defined, on a leaf—Phytotoxicity.

162 Round, beige spots on a leaf, browner at the periphery—Adverse climatic conditions.

MELON

- **Various phytotoxicities**

In certain circumstances the application of pesticides or fertilisers, whether intended or accidental, to cucurbits may cause localised burning of the foliar tissues, appearing as spots of various sizes. For example, we have observed this type of damage in the following cases:

○ After the wind has carried droplets of herbicide (sprayed near the crop) onto the plants (**159**);
○ Following treatment with an anti-mildew fungicide (**160**), or a mixture of pesticides (fungicide, insecticide, etc.) with a foliar fertiliser;
○ After broadcasting fertiliser onto the crop (**161**).

It is often only the most exposed leaves that are affected, e.g. those near the doorways and ventilators of glasshouses, etc., or those on one side only of the plants. In the latter case the symptoms are particularly obvious in those parts of the plant where residues of the product or products are most likely to remain.
Phytotoxicities of a different type, in other situations, can cause leaf spots on melon leaves and also on other cucurbits.

See also pages 34, 35, 62, 63.

- **Adverse climatic conditions**

Where condensation is likely to be heavy, in particular in low continuous tunnels, the numerous drops of water on the leaves act as magnifying glasses in very sunny weather. This causes burning of the underlying foliar tissues (**162**).

163 Grey-green circular patch, with yellow halo, on the edge of a cucumber leaf—*Didymella bryoniae*.

164 The spots starting to develop round the edge of the lamina are spreading towards the middle, forming brown/beige flame-shaped patches—*Didymella bryoniae*.

165 Grey-white or beige patch on the edge of a cucumber leaf—*Botrytis cinerea*.

166 Later, the patch becomes necrotic and shows darker concentric rings; the tissue becomes desiccated and splits—*Botrytis cinerea*.

167 Numerous beige spots with yellow haloes on the edge of a courgette leaf—*Botrytis cinerea*.

CUCUMBER
AND
COURGETTE

Spots with brown areas often at the edge of leaves

(sometimes showing concentric ring patterns)

POSSIBLE CAUSES

- *Botrytis cinerea* (factfile 7)
- *Didymella bryoniae* (factfile 8)
- *Pseudoperonospora cubensis* (factfile 11)

Marginal necrosis of the lamina (see page 111)

ADDITIONAL DIAGNOSTIC GUIDELINES

- *Botrytis cinerea, Didymella bryoniae*

These two fungi behave in the same way, especially with regard to protected crops of cucumbers, and very often they develop on the same plants and cause the same damage. They are opportunist organisms which frequently take advantage of particular conditions (excessive humidity under protection, stagnant water lying on the leaves particularly at the edges, numerous pruning wounds or a great deal of senescent tissue) to invade the plants and cause serious damage. Diagnosis is reasonably simple.

	Botrytis cinerea	*Didymella bryoniae*
Spots on edge of lamina	+	+ +
Fruit rot with pinched end	–	Brown to black + +
Fructification on affected parts	Grey mould = conidiophores and conidia (visible to the eye)	Black pits = pycnidia and/or perithecia (visible with lens)

Botrytis cinerea also invades other cucurbits, in particular the fruit (**354, 362, 367, 368**) and sometimes the leaves of courgette and melon plants. Generally, it develops first by colonising flower fragments (still clinging to the end of the fruit or fallen onto the fruit or the leaves). This source of nutrition is often essential, if the organism is to colonise successfully.

168 Brown wavy patterns and discrete grey mould are perfectly visible on this water melon leaf—*Botrytis cinerea*.

170 Canker on stem covered with grey mould—*Botrytis cinerea*.

169 Rot starting at the extremities of the fruit:
- Cover of grey mould (*Botrytis cinerea*, on left);
- Black, with pinched end (*Didymella bryoniae*, on right).

171 Area of oily discoloration on the stem, whose black colour is partly due to numerous fungal fructifications—*Didymella bryoniae*.

Although in France *Didymella bryoniae* is quite commonly found on cucumber crops grown under cover, it occurs far less often in the other cucurbits. In the melon, it causes more or less gummy cankers on the stem (**269, 272, 278**), and only occasionally spots on the leaves. It affects chiefly the fruit of the gourd causing rotting at the point of contamination and death of the seeds (**348**) and the seedlings that develop from these seeds (**172**).

- *Pseudoperonospora cubensis*

Cucumber leaves, especially those of plants grown under cover, are often covered by a film of water which allows cucumber mildew to develop rapidly in the form of large areas which later become desiccated. Often, other more characteristic spots can also be seen (refer to page 98).

172

172 Necrosis, originally oily then becoming greyish to brown on the cotyledons of a gourd, developing on the site of the seed-coat of a contaminated gourd seed—*Didymella bryoniae*.

173

173 Oily patches becoming beige to reddish-brown and desiccated, located particularly at the edge of the lamina of a cucumber leaf—*Pseudoperonospora cubensis*.

174 Round spots, sometimes angular, light green and slightly oily, on the upper side of a melon leaf—*Pseudoperonospora cubensis*.

175 On the underside, these spots appear at first to be very oily—*Pseudoperonospora cubensis*.

176 After the fungus has produced a greyish mauve velvety mass, the centre and then the whole of the spots quite rapidly become necrotic and desiccated—*Pseudoperonospora cubensis*.

177 Numerous characteristic brown spots; some leaves are partially necrotic—*Pseudoperonospora cubensis*.

MELON

The best time for easy examination of mildew fructifications is early in the morning, when ambient humidity is high and the conidia have not yet dispersed.

Oily spots on leaves

(subsequent yellowing and necrosis)

POSSIBLE CAUSES

- *Pseudomonas syringae* pv. *lachrymans* (see page 84)
- *Pseudoperonospora cubensis* (Downy mildew) (factfile 11)
- Melon Necrotic Spot Virus (see also pp 81, 82, 83) (factfile 28)
- 'Various physiological reactions'

ADDITIONAL DIAGNOSTIC GUIDELINES

During cultivation, melon and cucumber leaves are sometimes affected by oily patches which rapidly become necrotic. This oily look appears to be a common factor in many cases where the leaves suffer decay, and may have various causes:

- *Pseudoperonospora cubensis*

It is essential that an outbreak of this fungus is diagnosed as soon as possible because the resulting epidemics, like those of many downy mildews, spread extremely rapidly and are highly damaging. Fortunately they are easily identified even though the symptoms they produce (only on leaves) are quite variable (see also **173–190**). The fungus generally affects cucumbers and melons, but it could easily very soon become adapted to courgettes. In some areas it already affects courgettes, water melons and other cucurbits; growers should therefore keep a sharp lookout for this disease.

- Melon Necrotic Spot Virus (MNSV)

This virus, like the other viruses transmitted by *Olpidium*, initially produces very characteristic, small, oily, necrotic spots on melon (**183**) and cucumber leaves. The way the spots develop makes their identification relatively easy (refer to pages 81, 82, 83). Infection by these viruses tends to occur more often during the seasons when days are shorter and temperatures lower (early or late crops, particularly those grown under cover). They can cause severe damage in soil-less systems of cultivation.

178 Angular spots bounded by the veins, oily to pale yellow, then necrotic, on the upper surface of a cucumber leaf—*Pseudoperonospora cubensis*.

179 On the underside, these spots are oily to greasy; the vein boundary is clearly marked—*Pseudoperonospora cubensis*.

180 Sometimes the symptoms are located on one particular part of the leaf; here the oily symptom is seen around the leaf periphery—*Pseudoperonospora cubensis*.

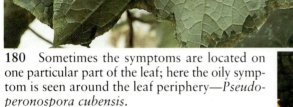

181 The outbreak may be more diffuse and cover the entire lamina which becomes pale yellow—*Pseudoperonospora cubensis*.

182 The discrete but sometimes numerous grey to mauve-brown fructifications seen here on the underside of the lamina tend to accumulate along the veins—*Pseudoperonospora cubensis*.

In addition to the two diseases described above there are others causing oily, greasy symptoms (often in the form of spots), at least initially. These occur, for example, with *Pseudoperonospora syringae* pv. *lachrymans*, on cucumbers (see page 84), with *Xanthomonas campestris* pv. *cucurbitae* on gourds (see page 85) and with *Ulocladium atrum* on cucumbers (see page 86). Remember that some phytotoxic responses (superficial leaf burn) result in symptoms which are initially oily in character. Apparently this type of symptom is the result of a specific behavioural factor in the plant, **a physiological reaction** to damage. This is shown very clearly by certain resistant varieties when subjected to the micro-organisms to which they are resistant. Oily and necrotic deterioration can be seen affecting a parent melon plant resistant to mildew and inoculated with this parasite (**184, 185**). Here, the defensive response of the host is shown by a localised hypersensitive reaction leading to these symptoms.

It is not always easy to determine the precise cause of oily spots; the reader is advised to proceed as follows:

- Search for the most characteristic symptoms when these exist;
- Carefully study the distribution of the disease on the plant and the crop;
- Finally, if there is any doubt, contact a diagnostic laboratory.

183   Small greasy spots, with rapidly necrosing centres—Melon Necrotic Spot Virus.

184

184   Melon leaf covered with small yellow spots, some necrotic—Physiological reaction (hypersensitivity).

185   On the underside, these spots appear oily and necrotic—Physiological reaction (hypersensitivity).

186 Yellow diffuse patches on the upper side of a melon leaf—*Sphaerotheca fuliginea.*

188 On the underside, a discrete white downy mould covers the patches—*Sphaerotheca fuliginea.*

190 Yellow spots on a cucumber leaf bounded by and located along the veins—*Pseudoperonospora cubensis.*

187 Discrete, sometimes angular, yellow spots visible on the upper side of a cucumber leaf—*Leveillula taurica.*

189 On the underside, the spots appear more angular and are covered with a discrete white downy mould—*Leveillula taurica.*

Yellow spots on leaves

POSSIBLE CAUSES

- *Erysiphe cichoracearum* (Powdery mildew) (factfile 9)
- *Leveillula taurica* (factfile 10)
- *Sphaerotheca fuliginea* (Powdery mildew) (factfile 9)
- *Pseudoperonospora cubensis* (Downy mildew) (factfile 11)

- Oedema

ADDITIONAL DIAGNOSTIC GUIDELINES

- **Powdery mildews affecting cucurbits (*Erysiphe cichoracearum* or *Sphaerotheca fuliginea*)**

When these organisms develop normally they produce very characteristic symptoms on the leaves of cultivated cucurbits (powdery white spots, see pages 102–103). Sometimes, when the host is only fairly susceptible, or not at all, yellow spots may appear on the leaves (**186–188**) on which these mildews apparently have some difficulty producing spores. This is also true for *Leveillula taurica*, another powdery mildew fungus with a host range; it sometimes occurs on cucumbers where it produces relatively discrete, often angular, yellow spots (**187–189**). On the underside of the leaves the spots may show a light covering of white conidia. The fungus is not commonly found on cucumbers and any damage caused is not normally serious. We have never seen it on melon or courgette, although it has been reported in the latter host.

- *Pseudoperonospora cubensis*

Cucumber downy mildew develops rapidly on the leaves. The oily spots which it produces (see pages 96–98) disappear quite quickly and frequently the observer sees only yellow spots (**190**) or alternatively dark green and light green/yellow patches giving the leaf a mottled appearance. In any event it is quite an easy disease to identify.

- Oedema

When humidity of covered crops is very high, spots may appear on the upper side of the leaves and then show a tendency to become necrotic (**191–192**). On the underside, it can be seen that these spots correspond to the 'islets' of cells which have ruptured as a result of the high humidity. Crops grown in plastic tunnels, low level continuous tunnels and Agryl P17, all non-woven materials which tend to maintain a very high ambient humidity, are particularly at risk.

191

191 Diffuse, chlorotic, necrotic spots on the upper side of a melon leaf—Oedema.

192 'Islets' of ruptured cells with centres becoming necrotic—Oedema.

192

193 Numerous small, circular, powdery spots on the upper side of a melon leaf—*Sphaerotheca fuliginea*.

194 Powdery white patches (the spots having joined together) on a cucumber leaf—*Erysiphe cichoracearum*.

195 Totally mildewed courgette leaf—*Sphaerotheca fuliginea*.

196 Water melon is normally not susceptible, except for one variety on which mildew creates the classic powdery white spots—*Sphaerotheca fuliginea*.

197 Mildew also colonise the underside of the leaves—*Sphaerotheca fuliginea*.

Mildew on cucurbits

Powdery white or buff spots on leaves

POSSIBLE CAUSES

- *Erysiphe cichoracearum* (factfile 9) } Powdery mildew
- *Sphaerotheca fuliginea* (factfile 9) }

- Deposits of pollen

ADDITIONAL DIAGNOSTIC GUIDELINES

- *Erysiphe cichoracearum, Sphaerotheca fuliginea*

These two fungi are the principal cucumber powdery mildews. They cause very characteristic symptoms which are exactly the same on the leaves, and also on the petioles and stems (**261**) but more rarely on the fruit (**338**). Diagnosis is very easy but it is impossible to differentiate between the different mildews with the naked eye. It may, however, sometimes be of advantage to distinguish one mildew from the other, because they are not equally susceptible to certain fungicides. The non-selective use of fungicides may explain why the latter are sometimes ineffective. There are some varieties, particularly of melon and cucumber, which are resistant to these organisms (see Appendix III).

Several saprophytic fungi of the leaf surface, and more rarely hyperparasites of mildews, live on mildew colonies or on the underlying damaged tissues: **Stephanoascus sp., Ampelomyces quisqualis, Tilletiopsis spp., Verticillium lecanii,** various **Penicillium spp., Cladosporium spp., Acremonium sp.**

Their presence on mildew colonies can be recognised by the following changes:
- To the colour of the spots which in places will become greyer or reddish-brown to olive-brown (**199**);
- To the appearance of the spots on which the mould of which they are formed becomes less homogeneous, and sometimes flattened.

High humidity favours their development.

199 The powdery white spots are covered by various fungi which are hyperparasites on mildews—*Erysiphe cichoracearum/Cladosporium* sp.

198 As well as the white mould, the mildewed leaves sometimes show minute black dots: the perithecia, sexual reproductive organs of mildew—*Sphaerotheca fuliginea*.

- Pollen deposits

Producers are sometimes worried by the appearance of fawn spots on the surface of the leaves (200) and fruit of melon plants (342). These spots are slightly raised and have a powdery appearance. If they are rubbed with a finger it can be seen that they are in fact superficial deposits which can easily be removed. There is no mystery about their origin: they are left by bees which are very busy round melon plants (which they pollinate) and whose legs are laden with pollen grains. These collect, forming small masses which, when they reach a certain size and if the weather is wet, can fall onto the melon leaves and fruit. The deposits are not harmful to the organs affected.

200

200 Ochre/fawn spots, slightly raised, on the upper side of a melon leaf—Pollen deposit.

Wilted, desiccated leaves

(may or may not be preceded or accompanied by yellowing)

POSSIBLE CAUSES

- *Pseudomonas marginalis*
- *Pseudomonas viridiflava*

- Diseases causing decay in roots, collar and stem
- *Pseudoperonospora cubensis* (factfile 11)

- Cucumber Mosaic Virus (**CMV**) (**on cucumbers**) (factfile 29)
- Zucchini Yellow Mosaic Virus (**ZYMV**) (factfile 32)

- Adverse climatic conditions
- 'Crown blight'
- Non-parasitic wilting of melon
- Physiological drying of melon
- Peripheral necrosis of the leaf
- Various phytotoxicities

HOW TO ANALYSE AND UNDERSTAND WILTING

Normally, when the leaves of one or several plants turn yellow and wilt, and/or also become desiccated, it is useless to look for the cause on the foliage itself. When confronted with these symptoms, growers and their staff often tend to focus their attention on the leaves expecting to find in them the cause of the trouble. Often they will take only the leaves to the plant laboratory whose technicians will then find it impossible to diagnose the source of the problem. As a general rule, the origin of wilting and desiccation (the last stage of wilting) should be sought elsewhere than in the leaves: i.e., in particular in the roots, the collar, and the exterior or interior of the stem.

Various forms of deterioration occurring in these organs disturb the functions of water and mineral absorption (roots) and their transport (collar, stem). The most obvious result is that the plants lack water and wilt either temporarily or permanently and become partially or totally desiccated.

When and how rapidly wilting will develop depends mainly on three parameters:

- the stage of development of the plants; plants laden with fruit and approaching harvest wilt very easily;
- the nature and severity of the deterioration at its source; in the early stages, before too much damage has been done, the wilting may be reversible; at night the plants lose less moisture through evaporation and may regain turgor.
- climatic conditions; high temperatures, powerful forced ventilation, a strong wind, by increasing evapotranspiration by the plants all encourage early wilting. Conversely, in conditions where evapotranspiration is low, e.g. in a cold, wet winter, disease in the plants (canker in the collar, root rot etc.) will not become evident until deterioration is already advanced; by the time it is discovered, it is often too late to intervene.

In some cases, wilting may occur in a particular part of the leaves and plants, characteristic of one or several diseases. You can find out about this later, particularly after looking at the vascular diseases.

Leaves are a good indicator of the plant's water supply. Wilting or desiccation can be considered as being the consequences of disturbances whose origins must be detected by careful examination of all the organs which could be responsible.

Examples of wilting and desiccation of leaves

201 Melon plants on which numerous leaves are wilted or desiccated—*Phomopsis sclerotioides*.

202 Cucumber leaf in the course of wilting—Cucumber Mosaic Virus (CMV).

203 Desiccation of the inter-veinal tissues of a melon leaf—*Pyrenochaeta lycopersici*.

204 Wilting of a melon leaf preceded or accompanied by yellowing of the lamina—*Phomopsis sclerotioides*.

When you see yellowing and wilting of the leaves act at once!

Also examine

The stem (exterior and interior)

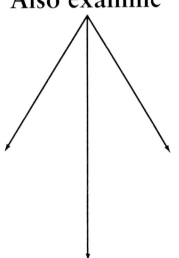

The roots

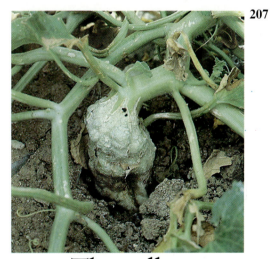

The collar

209

209 Melon leaves totally desiccated, still attached to the plants in an erect but curled position—*Pseudoperonospora cubensis*.

210

210 Brown inter-veinal necrosis visible on the upper side of a melon leaf—Physiological drying of melon.

211

211 The leaves are rapidly becoming necrotic, desiccated and wilted—Physiological drying of melon.

212

212 Inter-veinal necrosis and desiccation of the lamina of a cucumber leaf, following burning—Various phytotoxicities.

- **Adverse climatic conditions**

Field grown cucurbits are very vulnerable to adverse climatic conditions. These are a basic cause of leaf damage resulting in very similar symptoms whatever the environmental factors. Such damage has both long-term and short-term effects on the crop. It often occurs when the weather is cold (**213**), stormy, very windy (**214**) or when wet leaves are exposed to hot sunshine. In the latter case, the droplets lying on the lamina act as a magnifying glass for the sun's rays, burning the underlying leaf tissues. Similarly, in covered cultivation where white plastic film is often used, particularly as a surface mulch, inter-veinal burning can be seen on cucumber leaves close to the soil (**215**), the result of sunlight reflecting back from the film.

- **Peripheral necrosis of the lamina**

The periphery of the lamina will often be affected by yellowing and more or less serious necrosis. We have pointed out that among the possible causes are **phytotoxicites** and **excess salinity** (see page 64). There are of course others. When there is high atmospheric humidity water accumulates on the edges of the leaves and if it remains there for some time, it can cause slight local damage. More often, this damage is caused by contamination of the water with pesticides or various bacteria, such as *Pseudomonas marginalis* and *Pseudomonas viridiflava*. These micro-organisms, reported to infect cucumbers in Japan, are not dangerous pathogens and usually the deterioration they cause has not harmful effect on the crop.

- **Crown blight**

The very high levels of resistance to mildew shown by both melons and cucumbers are often associated with necrosis (**216–217**), which affects the plants particularly during seasons when days are short and light is restricted (spring and autumn). Very early crop production using varieties with a very high mildew resistance level may therefore cause problems. This factor has delayed the development of new varieties resistant to *Oidium* spp.

216 Primary, beige, dry-looking inter-veinal necrosis, visible on both sides of a melon leaf—Crown blight.

217 Necrosis has developed causing extensive interveinal desiccation—Crown blight.

- **Cucumber Mosaic Virus (CMV)**

In addition to the normal symptoms of mosaic, this virus can cause sudden generalised wilting of cucumber plants, depending on climatic conditions and the varieties or hybrids; the basic cause of this effect is not yet known (**218**). There are other symptoms which will help you to make a reliable diagnosis: refer to pages 25-29-49.

- **Zucchini Yellow Mosaic Virus**

Certain strains of this virus (pathotype F) can cause rapid wilting (**219**) followed by the total desiccation of the plants in varieties such as Doublon or Alpha which carry the gene *Fn* (*Flaccida necrosis*) (refer to the factfile ZYMV and Appendix III).

218

218 Wilting and yellowing of cucumber whose fruits are slightly mottled—Cucumber Mosaic Virus (CMV).

219 The top of a melon plant wilting and necrotic; a superficial reddish-brown necrosis may develop on part of the stem and on the petioles—Zucchini Yellow Mosaic Virus (ZYMV).

219

Abnormalities and decay in roots and/or collar and/or stem base

SYMPTOMS EXAMINED

- Yellowing, browning, suberisation of the roots and/or the collar and/or the stem base
- Various types of decay of the collar and/or the stem base
- Various types of decay in the roots, collar and stem base

POSSIBLE AND PROBABLE CAUSES

- *Agrobacterium rhizogenes*
- *Erwinia* spp.
- Actinomycetes
- *Chalara elegans*
- *Corticium solani*
- *Didymella bryoniae*
- *Fusarium* spp.
- *Macrophomina phaseoli*
- *Monographella cucumerina*
- *Monosporascus cannonbalus*
- *Monosporascus eutypoides*
- *Myrothecium roridum*
- *Penicillium oxalicum*
- *Phomopsis sclerotioides*
- *Phytophthora* spp. and *Pythium* spp. = various Pythiaceae
- *Pyrenochaeta lycopersici*
- *Rhizoctonia solani*
- Various fungi
- *Meloidogyne* spp.
- Root asphyxia and/or soil too cold
- Physiological rupture
- Excess salinity
- Phenomenon of root regression in cucumbers
- Various phytotoxicities

We have grouped the decay in roots, collar and stem base in the same section and under the same heading on purpose, for the following reasons:

- It is impossible to separate these three plant sections, because very often fungi attacking the roots (most common location) are also able to colonise the collar and the stem base. The reader who encounters a problem affecting these parts of the plant should consider them as a whole;
- There is enormous difficulty in diagnosing the diseases which affect these various plant parts in cucurbits because their roots are particularly susceptible to attack by both parasites and non-parasites. Another problem is that most diseases which affect the roots cause almost exactly the same symptoms. The risk of confusion is therefore very high and it is not possible to offer a simple, structured diagnostic approach, such as you will find in the other sections.

OBSERVATION GUIDE

**PULL OUT THE ROOTS CAREFULLY AND
WASH THEM THOROUGHLY**

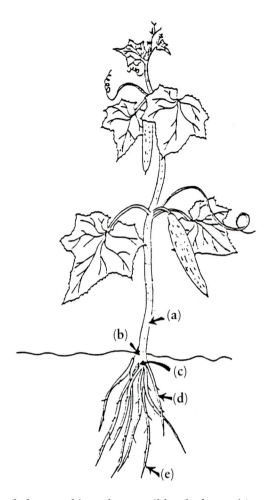

Locality of the main parts below, and just above, soil level of cucurbits:

(**a**) Stem base
(**b**) Collar
(**c**) Main root (the taproot)
(**d**) Secondary root
(**e**) Radicle

VERY DIFFERENT DIAGNOSIS

(Follow the correct procedures to obtain a more accurate diagnosis.)

The root system is the least well understood part of the plant; many growers and their staff do not know how to assess its condition, because often they do not go the correct way about it.

First, it is advisable to **pull up the roots carefully;** avoid dragging them out roughly or the diseased parts, the most fragile (but also the most important from the diagnostic aspect), will remain in the ground. Secondly, **they should be washed thoroughly in water** to remove particles of earth which may be masking certain symptoms.

Now you can examine the roots: do this very carefully, using a magnifying glass if you have one.

If the collar and the stem base are affected, examine the entire root system also, referring to the section 'External/Internal Abnormalities and Decay of the Stem'.

Examples of abnormalities and decay of the roots and collar

220 Brown, rotted cucumber root—*Phomopsis sclerotioides*.

221 Corky cucumber roots, with partially decomposed cortex; black dots are also visible—*Phomopsis sclerotioides*.

222 Cucumber roots ringed with galls—*Meloidogyne* sp.

223 Collar of gourd, necrotic and 'pinched'—*Phytophthora* sp.

224 Roots and collars of cucumber plants totally brown and necrotic—*Rhizoctonia solani*.

225 Numerous healthy roots coloured white; some are beginning to turn yellow—Root regression in cucumbers.

227 As well as the roots, the vessels (V) at the level of the main root and just above sometimes show more or less intense brown markings—Various Pythiaceae.

226 Brown, rotted radicles and roots—Various Pythiaceae.

228 Browning and partial destruction of the root cortex (E); the stele (C) is visible in places—*Phomopsis sclerotioides*.

229 Almost completely rotted root cortex—*Phomopsis sclerotioides*.

Yellowing, browning, corkiness of the roots and/or the collar and/or the stem base

POSSIBLE CAUSES

- *Erwinia* spp.
- *Didymella bryoniae* (factfile 8)
- *Fusarium* spp.
- *Phomopsis sclerotioides* (factfile 16)
- Various Pythiaceae (*Pythium* spp., *Phytophthora* spp. (factfile 14)
- *Pyrenochaeta lycopersici* (factfile 17)
- *Rhizoctonia solani* (factfile 15)
- Various fungi

- Asphyxia of the root and/or collar
- Excess salinity
- Non-parasitic wilting of melon (see page 109)
- Root regression in cucumbers

All diseases which affect the roots of cucurbits cause diffuse yellowing and/or browning (localised or general), necrosis, and the destruction of numerous roots. In the most serious cases, the entire root system may deteriorate, the vessels at taproot and collar level becoming yellow and brown (**227**). Decay may spread further to include the collar and stem base. Certain fungi frequently attack the collar and/or the buried parts of the stem before invading the roots.

Some fungi (these are rare) also cause very characteristic and easily identifiable deterioration in the roots which will be described in the following pages.

Identification of the other diseases is very difficult if it is based solely on the symptoms mentioned above; **laboratory investigation is essential** to determine the cause or causes. Generally it is fungi which are responsible, for example *Rhizoctonia solani* and the Pythiaceae (several species of *Phytophthora* and *Pythium*), or faulty cultural practices. The latter may be the cause of **root asphyxia**, or **root burns**, due to too high a concentration of salts (**excess salinity** resulting from excessive dressings of fertiliser or from drying out of the substrate).

These problems often occur simultaneously or, more accurately, some of them create the conditions favourable for others. For example, outbreaks of Pythiaceae are more severe if the plants have been over-watered; on the other hand, *Rhizoctonia solani* may be particularly active in soils which are too dry. In some cases, the roots may carry a large concentration of cysts and sporangia of *Olpidium* spp. Are these virus vectors acting as parasites on the roots? Are they biological markers indicating a root system in poor health?

One of the characteristics of cucurbit roots, especially in the melon, is their **tendency to become corky** when exposed to specific environmental conditions (excess water, soil too cold, excess salinity). Often growers mistakenly attribute this symptom to the ravages of *Pyrenochaeta lycopersici* (responsible for corky root disease).

If you consider the questions below carefully you may sometimes be able to find an explanation for your problem (an affirmative reply to any of the questions will confirm it):

- Have you over-watered the plants, or irrigated with water at too low a temperature?
- Are the diseased plants located in the wettest part of the growing area?
- Did you carry out planting at a time when the soil was still cold and wet?
- Did you over-water in the nursery or when planting out?
- Is your soil poorly drained?
- Did you apply too much fertiliser before planting out or during cultivation?

(If possible, consult a specialist laboratory.)

ADDITIONAL GUIDELINES

	Pythiaceae	*Rhizoctonia solani*	*Pyrenochaeta lycopersici*
Location of symptoms:			
• roots	+ +	+	+
• collar	+	+	−/+
• stem base	+/−	+/−	−
Visible signs of the pathogen on affected parts	None	Discrete brown mycelial filaments (**237**)	None
Stage symptoms observed:			
• nursery, after planting	+ +	+	
• adult	+	+	
Susceptible hosts (in decreasing order)	cucumber, melon, courgette	melon, cucumber courgette, water melon	melon
Frequency in crops	+ + +	+ +	+/−
Type of crop affected:			
• under cover	+ +	+	+
• field grown	+	+	+
• in soil	+	+	+
• soil-less	+ +	−/+	−

• **Root regression in cucumbers**

'Root loss' in cucumbers (browning, rotting and root disappearance) is quite commonly seen, in both soil-grown and soil-less cultivation. In the latter case, the result is quite spectacular as the easily visible roots (**206, 225**) seem to change very rapidly. The cause or causes are not thoroughly understood; there appears to be several contributory factors:

• The number of fruit per plant: heavy-cropping plants seem to be more susceptible. There may be competition between the fruit and the roots for the plant's assimilated food supplies at the expense of the roots;
• The prolific development of the first fruits on the stem: plants on which many fruits set too early will limit their root development;
• Too much water, the efficiency of oxygenation of the root environment: asphyxiation encourages root loss;
• The plant's food supply and the climatic conditions (temperature of the substrate in summer): these two factors are not well understood;
• The presence of *Pythium* spp. (see page 120); their role has not yet been fully clarified: are they always parasitic fungi (in view of the pathogenicity of some of them this may be possible)? Do they intensify a physiological phenomenon? Current studies are likely to provide us with an answer in the near future.

Phomopsis sclerotioides	*Didymella bryoniae*	*Penicillium oxalicum*	*Fusarium* spp.
+	−/+	−	+/−
+/−	+	−	+
−	+	+	+
'Black lines', small black dots (**241, 243, 244**)	Tiny black dots (photo **247**)	Grey-blue mould (photo **248**)	Whitish to pink mould (photo **249**)
−	+/−	−	+
+	+	+	+
Cucumber, melon	Cucumber, melon	Cucumber	Cucumber, courgette, melon
+	+	+	+/−
+ +	+	+	+
+	−/+	−	+
+ +	+	+	+
+/−	+	+	+

230   Browning and rotting of roots in a young cucumber plant.

231   Dark brown/black moist lesion, diffuse at its encroaching edge, having crept above the collar of the melon plant.

232   Note also the deterioration in depth of the tissues, and in particular the yellowing of part of the vessels.

233   Generalised rotting of the buried part of a cucumber; only vascular tissues are left at the level of the collar.

Various Pythiaceae (*Pythium* spp., *Phytophthora* spp.)

- **Various Pythiaceae**

The Pythiaceae include fungi belonging to the genera *Pythium* and *Phytophthora*. They have some characteristics in common:

- They are perfectly adapted to an aquatic life and flourish in wet soils or substrates with high water retention. Some species will grow at an enormous rate in soil-less, and especially hydroponic (NFT), systems.
- They are not host specific and are thus capable of invading a variety of hosts, including many vegetable crops. Almost all species and varieties of cucurbits are susceptible, especially the cucumber and, to a lesser degree, the melon and the courgette;
- Damage is particularly severe in the early stages of the seedling growth. In the nursery, they cause damping-off and losses among the emerging seedlings. The same damage occurs where direct drilling is practised (see Appendix I). It is advisable to take care when pricking-out and planting. Adult plants may also be affected, especially in soil-less systems; the resulting root loss causes great damage to the cucumber crop;
- The symptoms caused in the collar and roots and those at the stem base are very similar, such as root rot (**230**), and cankers and rotting of the collar (**231**). When rainfall is heavy or after copious watering the parts above ground become splashed with dirty water, and the organisms may invade them and cause leaf spots (**129–130**), necrosis of the apex (**131**) and fruit rot (**132**). In the same way, some parts which are above ground level (stem, fruit) but are in contact with wet soil may also be contaminated and suffer damage (**262**). Melons, courgettes and gourd plants are particularly susceptible to *Phytophthora capsici*.

Several species of *Pythium* and *Phytophthora* can attack cucurbits; those most commonly reported are:

- *Pythium aphanidermatum*
- *Phythium de baryanum*   } the most common
- *Pythium ultimum*

- *Pythium butleri*
- *Pythium intermedium*
- *Pythium irregulare*
- *Pythium splendens*

- *Phytophthora capsici*
- *Phytophthora cryptogea*
- *Phytophthora drechsleri*
- *Phytophthora melonis*

Phytophthora melonis is able to invade all plant parts of cucurbits, but especially cucumbers where it causes considerable damage as a collar rot pathogen. In China and Japan it is known to be a major cause of rot, and other species, *Phytophthora sinensis,* cause seedling and adult plant death in cucumbers.

234 Partially necrotic cucumber roots; note the disappearance of numerous lateral roots and the browning of large sections of the roots—*Rhizoctonia solani.*

235

235 Several cucumber roots are superficially cracked and corky; in places they are brown and rotted—*Rhizoctonia solani.*

236

236 Like the roots, the taproot and the collar of these young melon plants are brown and corky with superficial rotting—*Rhizoctonia solani.*

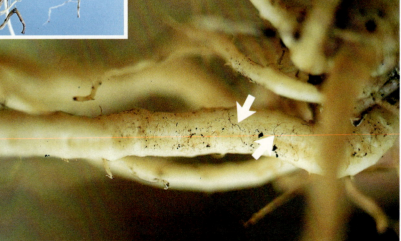

237

237 Fine, brown mycelial filaments, very characteristic of the fungus, can be seen on the surface of the roots—*Rhizoctonia solani.*

Rhizoctonia solani

● *Rhizoctonia solani*

This highly polyphagous fungus, present in virtually all market garden soils, is found as often in nurseries on young plants (see Appendix I) as it is on adult plants. Too often underestimated, it can produce serious damage in adult plants:

● On the roots and collar it causes more or less marked browning (**234–236**); roots may appear corky and/or rotted (**235**);
● The other organs are sometimes affected, in particular the fruit (**344**) and the stems (**263**) in contact with the soil.
   Brown mycelial filaments appear on all the infested organs, and are very characteristic (**237**).

● *Pyrenochaeta lycopersici*

As on the tomato (its preferred host), this fungus causes more or less extensive patches of corky deterioration in melon roots (**238**). As we have pointed out previously, melon roots have a natural tendency to a greater or lesser degree of suberisation when conditions are unfavourable. This is what happened to the melon stem base shown in **239**, when it was planted too early in cold, wet soil; the buried section was affected by these adverse conditions.

239 Collar of melon plant with a corky appearance, wet to the touch, showing longitudinal cracks; the roots are in the same condition—Soil too wet and cold at planting.

239

● *Pyrenochaeta lycopersici*

238 Dark reddish-brown corky sleeve, showing longitudinal cracks in places—*Pyrenochaeta lycopersici*.

238

240 Large cucumber root visible on the soil surface; where these are numerous, such roots give an 'oak tree foot' appearance to the stem base—*Phomopsis sclerotioides*.

242 Very swollen collar; nearly all the roots have rotted and disappeared—*Phomopsis sclerotioides*.

Phomopsis sclerotioides

241 Pale brown, rotted cucumber root; black lines can be seen on the cortex—*Phomopsis sclerotioides*.

243 In addition to the black lines, parts of the roots are covered with an enormous number of small black dots (the pseudosclerotia of the fungus)—*Phomopsis sclerotioides*.

244 Sclerotic black patches are visible in places beneath the cortex—*Phomopsis sclerotioides*.

● *Phomopsis sclerotoides*

This soil fungus causes a great deal of damage to cucumbers and it also affects melons and the melon rootstock *Benincasa cerifera*. It causes a more or less wet rot on the roots which later partially disappear. It is difficult to identify, although the root symptoms are very characteristic (**240, 241, 242, 243, 244**). Pull up some roots very carefully and examine them closely.

Many other parasitic micro-organisms affecting the roots and collar of cucurbits have been reported from various countries:

- *Chalara elegans (Thielaviopsis basicola)*
- *Corticium rolfsii (Sclerotium rolfsii)*
- *Erwinia* spp.
- *Macrophomina phaseoli*
- *Myrothecium roridum*
- *Monographella cucumerina* (= *Microdochium tabacinum*)
- *Monosporascus eutypoides*

Their symptoms are often very similar and generally involve root rot. The only method of identifying them is to carry out microbiological tests on the affected plant parts.

● *Erwinia* spp.

When humidity is very high (heavy rain, temporary flooding of the plot, etc.) the plant, especially courgettes, may develop open cankers on the collar and sometimes the stem. When the stem is cut open the inside will appear more or less liquefied and very often will give off a disgusting stench (**245**). Various *Erwinia* species, especially *Erwinia carotovora* var. *carotovora*, are responsible for this type of damage. The bacteria penetrate the plants via petiolar wounds and rapidly affect the stem.

There can be another cause for this phenomenon. The split in the stem may be physiological in origin (similar to growth splits in fruit), and occur after heavy rainfall; *Erwinia* is then a secondary cause. This problem is very rare in France.

245

245 Rot has completely hollowed out the inside of this courgette stem—*Erwinia carotovora* var. *carotovora*.

246 Dark green, moist decaying tissues, gradually becoming covered with black dots (perithecia and pycnidia) which will in time colour the damaged tissues dark grey/black—*Didymella bryoniae*.

247 Decay may spread several centimetres above the collar, as in this cucumber stem base where the diseased tissues have become dark grey/black—*Didymella bryoniae*.

248 Decayed tissues are wet and covered with a grey-blue mould, developing on splits often found on the stem base of the cucumber—*Penicillium oxalicum*.

249 Comparatively dry canker encroaching gradually on the stem base of a cucumber, covered by pinkish mould—*Fusarium* sp.

Various types of decay in collar and/or stem base

POSSIBLE CAUSES

- *Didymella bryoniae* (factfile 8)
- *Fusarium* spp.
- *Penicillium oxalicum* (factfile 19)

ADDITIONAL DIAGNOSTIC GUIDELINES

- *Didymella bryoniae*

A fungus which is principally parasitic on the parts of cucurbits growing above ground level. It is sometimes found in the parts which are buried just beneath the soil surface;

- In Denmark, it causes a black discoloration of the roots of cucumbers grown under protection, because of the presence of numerous fungal fructifications in the cortex;
- It sometimes affects the collar of melon plants (**246**) and gourds; in Italy, it causes sticky cankers on gourds. In Northern Europe it is quite frequently found in the stem base of cucumber plants (**247**).

- *Penicillium oxalicum* (= *Penicillium crustosum*)

This fungus on cucumber pathogen has only recently been reported, its development in Europe dating from around 1987. It affects mostly the upper parts of the stem (**273**) but in certain conditions it penetrates through the small splits which often occur on the stem close to ground level. Once it has gained entry through these natural wounds it invades the tissues and causes a wet rot (**248**). In serious cases only the woody fibres remain. The grey-blue mould which appears on the damaged tissues is very characteristic of this fungus.

- *Fusarium* spp.

In France and the UK, the non-vascular *Fusarium* spp. are comparatively rarely found on cucurbits. Where decay occurs in the stem base of cucumbers grown under protection (**249**), it is likely that the cause is one or other of the weak *Fusarium* spp. which develop on plants which are already in bad condition for reasons which are not fully understood but are not the result of a pathogen. These *Fusarium* differ from strain 1 of **Fusarium solani** f.sp. **cucurbitae** which prefers to attack the collar of cucurbits (including *Cucurbita ficifolia*, the rootstock of the cucumber) and sometimes melon and cucumber. In France and the UK this fungus is very rare. We have several times isolated it from decayed collars on courgette plants, sometimes in association with *Rhizoctonia solani*. In Holland and in Israel, a fungus named there as **Fusarium javanicum** has also been reported on courgettes.
Strain 2 of *Fusarium solani* f.sp. *cucurbitae* invades the fruit only.

250 White galls visible on cucumber roots like a string of beads—*Meloidogyne* sp.

251 The galls often have a tendency to turn brown and become slightly corky—*Meloidogyne* sp.

252 Close to soil level, a corky split several centimetres long can be seen in this stem—Physiological split.

253 Grossly swollen collar: the adventitious roots have been severely suppressed—Phytotoxicity.

Various abnormalities of the roots, collar and stem base

POSSIBLE CAUSES

- Actinomycetes
- *Agrobacterium rhizogenes*

- *Fuligo septica*
- *Meloidogyne* spp. (factfile 21)

- Physiological splitting
- Various phytotoxicities

ADDITIONAL DIAGNOSTIC GUIDELINES

- *Meloidogyne* spp.

The roots of cucurbits are often affected by galls, which turn from pearly white to brown as a result of superficial corkiness (250–251). Plants in this family are particularly susceptible to gall nematodes which are found in soils where cucurbits or vegetable crops (Solanaceae, Compositae, etc.) have already been grown for several seasons. The highly characteristic galls make diagnosis very simple.

In Japan, actinomycetes also cause pale brown tumours on cucumber roots.
In England, a condition is reported where cucumber roots grow upwards on the soil surface ('Rhizomania'). The bacterium responsible for this curious phenomenon, *Agrobacterium rhizogenes*, not only causes the roots to grow upwards, but causes a reduction in plant growth and reduced yield.

- Physiological splitting

The stem base of the cucumber has a tendency to split longitudinally (252). This is very often the result of an over-vigorous response by the plant to stress, especially climatic stress. Normally these splits are infrequent and cause no damage as the tissues rapidly form protective scar tissue.

- Various phytotoxicities

We have seen many field-grown melon plants affected by a swelling of the collar (253); this is caused by the unsuitable use of herbicides belonging to the toluidine group (pendimethalin). The same symptom is apparent in similar circumstances in other crops such as the tomato. The effect on melons is quite striking, because this species is very susceptible to herbicides. In the present case, pendimethalin inhibits development of the lateral roots; swelling is due to the aborted development of numerous roots.

- *Fuligo septica*

In England, one of the myxomycetes develops round cucumber stems at just above soil level, giving them a bizarre appearance. The initially reddish-brown, viscous mass produced by *Fuligo septica* causes no adverse reaction in the plant and should be considered solely as a curiosity.

External/internal abnormalities and decay of the stem

SYMPTOMS EXAMINED

- Abnormalities and decay developing in any part of the stem
- Decay frequently beginning in pruning wounds and senescent tissues on the stem
- Internal decay of the stem

POSSIBLE CAUSES

- *Erwinia carotovora* var. *carotovora*
- *Erwinia tracheiphila*
- *Botrytis cinerea*
- *Cladosporium cucumerinum*
- *Colletotrichum lagenarium*
- *Didymella bryoniae*
- *Erysiphe cichoracearum*
- *Fusarium oxysporum* f.sp. *cucumerinum*
- *Fusarium oxysporum* f.sp. *melonis*
- *Fusarium oxysporum* f.sp. *niveum*
- *Fusarium* spp.
- *Macrophomina phaseoli*
- *Penicillium oxalicum* (= *Penicillium crustosum*)
- *Phytophthora capsici*
- *Rhizoctonia solani*
- *Sclerotinia sclerotiorum*
- *Sphaerotheca fuliginea*
- *Verticillium dahliae, Verticillium albo-atrum*

- Melon Necrotic Spot Virus
- Various viruses

- Hailstone damage
- Fasciation of the stem

We would also advise you to consult the section 'Abnormalities and Decay in the stem base'.

OBSERVATION GUIDE

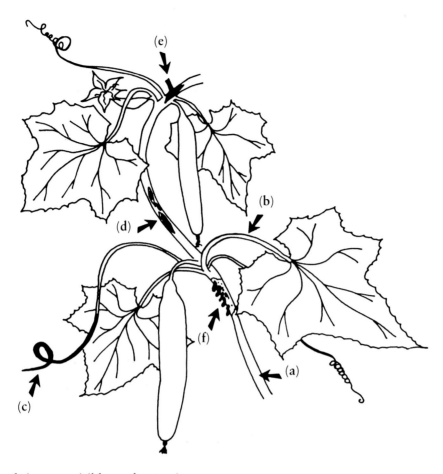

What external signs are visible on the stem?

(a) Stem
(b) Petiole
(c) Decay affecting the tip of a tendril
(d) Elliptical spots on the stem
(e) Canker becoming established in a wound caused by thinning out or disbudding
(f) Gummy canker on stem

Gummy exudates, in varying amounts, can often be found on the stems, peduncles and fruits of cucurbits. They are light coloured initially, becoming brown later through atmospheric oxidation. Known as **gummosis**, this condition is often associated with fusarium wilt of melon and is quite a common symptom found with other diseases, in particular:
- Fusarium wilt diseases, affecting especially melon, and to a lesser degree cucumber and water melon plants (**287, 296, 302**);
- *Didymella bryoniae*, which causes a gummy canker on melon stems (**269**).

Gummy exudate may also be present on lesions caused by several pathogenic micro-oganisms such as *Sclerotinia sclerotiorum*; the exudates are not characteristic of these diseases. They can also be found on the sites of various wounds of mechanical origin where the sap has leaked out. In other words, once a plant has suffered damage or decay of some sort, you may find these gummy globules at the site in question.

During cultivation, cucurbit stems develop some particularly vulnerable sites such as senescent tendrils, stubs of petioles, or peduncles left on the plants after thinning-out or harvesting. These sites, acting as 'nutrient bases', favour the development of several opportunist fungi, which are the causes of often serious damage.

254 Dark reddish-brown decay covered by a grey felting at the tip of a water melon stem—*Botrytis cinerea*.

256 Brown, wet rot on a section of a courgette petiole—*Phytophthora capsici*.

255 Small, elongated, beige cankers on a bottle gourd—*Colletotrichum lagenarium*.

257 More/less marked browning of the vessels, visible in the interior of a cucumber stem—*Fusarium oxysporum* f.sp. *cucumerinum*.

258

259

260

258 Melon stem showing pale reddish-brown elliptical cankers, with a velvety dark green central discoloration—*Cladosporium cucumerinum*.

259 Small, elongated cankers, moist at the beginning of development, on a cucumber stem—*Colletotrichum lagenarium*.

260 Stem of water melon showing brown decay, with a scattering of small salmon-pink fungal masses—*Colletotrichum lagenarium*.

261

262

263

261 Numerous small, white, powdery spots on a cucumber stem—*Sphaerotheca fuliginea*.

262 Oily, sticky decay ringing the stem of a Provençal musky gourd—*Phytophthora capsici*.

263 Reddish decay, more/less superficial and spreading, on a water melon stem—*Rhizoctonia solani*.

Abnormalities and decay affecting any part of the stem and petioles

POSSIBLE CAUSES

- *Cladosporium cucumerinum* (factfile 5)
- *Colletotrichum lagenarium* (factfile 6)
- *Erysiphe cichoracearum* (factfile 9)
- *Macrophomina phaseoli*
- *Rhizoctonia solani* (factfile 15)
- *Sphaerotheca fuliginea* (factfile 9)
- *Phytophthora capsici* (factfile 14)

- Zucchini Yellow Mosaic Virus (ZYMV) (factfile 32)

- Damage from hailstones
- Stem fasciation

ADDITIONAL DIAGNOSTIC GUIDELINES

	Cladosporium cucumerinum	*Colletotrichum lagenarium*	*Phytophthora capsici*	*Rhizoctonia solani*
Characteristic structures	Dark green mould	Tiny salmon-pink masses	White mould +/− creamy	Discrete brown mycelial filaments
Hosts in Europe	melon, courgette	melon	melon, courgette, gourd	melon, water melon
Symptoms on other organs:				
• leaves	(112) +	(114) +	(129–130) +/−	−
• fruit	(118) +	(120) +	(132) +	(344) +
• collar	−	−	+/−	(236) +
• roots	−	−	+/−	(234–237) +
Origin of contamination	aerial	aerial	soil contact, in irrigation channels	soil contact

Refer to page 145, because certain vascular fungal pathogens cause external stem decay as well as a characteristic stem browning.

- *Macrophomina phaseoli*

This fungus causes decay, often initially greasy, in the stem and collar of melon plants. Amber-coloured globules may appear on the surface. Later these become desiccated and may be covered by innumerable black microsclerotia.

264

264 Superficial brown necrosis, so-called 'streaks', on the petiole and at the apex of a melon plant—Zucchini Yellow Mosaic Virus (ZYMV).

265

265 Part of a cucumber stem, flat and striated—Stem fasciation.

266

266 Small, corky, greyish splits, more/less superficial, on a melon stem—Hail damage.

- **Various viruses**

Certain viruses (e.g. Zucchini Yellow Mosaic Virus ZYMV) may, in addition to other symptoms (see pages 27–41), be responsible for the appearance of more or less extensive superficial brown areas (called 'streaks') on the stems of melon plants in particular (**264**). They may also be found at the apex of the plant and on the petioles. Sometimes the entire plant or complete branches are necrotic and desiccated (**213**).

- **Stem fasciation**

The stems of affected melon and cucumber plants appear to be completely flat, becoming gradually wider and wider towards the apex; they have a striated appearance due to the fusing of the secondary stems (**265**). This condition, encouraged by low temperatures, is extremely rare and affects only very few plants at one time, irrespective of variety. Some mutants carry the gene and transmit it to their descendants, all of which will be affected.

- **Damage from hailstones**

After heavy rain and hailstorms it is not uncommon to see small or large splits in the most exposed parts of the stems and peduncles, located at the points where the impact of hailstones has been heaviest (**265**). Fruit is often badly damaged (**340–341**).

267 Wet, beige canker developing at the junction of a tendril, where a melon fruit has been nipped in the bud—*Sclerotinia sclerotiorum*.

268 Later, the canker becomes covered by a dense white felting; brown, gummy globules may be visible and black masses, the sclerotia—*Sclerotinia sclerotiorum*.

269 The decay here is initially wet, then dark reddish-brown, affecting quite a large portion of the stem. Note the large amounts of gummy exudates and the tiny black dots (perithecia or pycnidia) on the stem—*Didymella bryoniae*.

270 Dark reddish-brown to brown decay, relatively superficial, developing at the site of a pruning cut on a melon stem—Melon Necrotic Spot Virus (MNSV).

MELON

Decay arising from pruning cuts and senescent tissues on the stem

POSSIBLE CAUSES

- *Botrytis cinerea* (factfile 7)
- *Didymella bryoniae* (factfile 8)
- *Fusarium* spp.
- *Macrophomina phaseoli*
- *Penicillium oxalicum* (factfile 19)
- *Sclerotinia sclerotiorum* (factfile 20)
- *Phytophthora capsici* (factfile 14)

- Melon Necrotic Spot Virus (**MNSV**) (factfile 28)

ADDITIONAL DIAGNOSTIC GUIDELINES

MELON

	Sclerotinia sclerotiorum	*Didymella bryoniae*	MNSV
Characteristic symptoms	Dense white mycelium large black sclerotia on surface rarely in the pith	Sticky brown exudates, abundant; brown speckling (perithecia or pycnidia)	None
Symptoms on other organs	Fruit (**355**)	Leaves, fruit (**116**)	Leaves (**138**)
Crops most affected	Chiefly protected	Protected & field grown	Protected. Rare in field grown.

- Melon Necrotic Spot Virus (**MNSV**)

This virus, transmitted by a soil fungus in the genus *Olpidium*, can also infect plants during pruning operations. It invades the plant producing streaks (areas of superficial brown discoloration) developing mainly at the site of pruning cuts (**270**), and highly characteristic small necrotic spots on the leaves. The information on pages 75, 81 and 82 will help you to identify this disease correctly.

271 Section of a stem, beige to dark green in colour, covered by a grey to dark reddish-brown mould—*Botrytis cinerea*.

273 Dark green to reddish-brown decay, with a central covering of mould, initially white turning to grey-green—*Penicillium oxalicum*.

272 Dry, beige canker (initially moist and dark green), covered with numerous black specks (= pycnidia or perithecia)—*Didymella bryoniae*.

274 White downy mycelia covering a canker which has developed from senescent tissues—*Sclerotinia sclerotiorum*.

CUCUMBER

	Botrytis cinerea	*Didymella bryoniae*	*Sclerotinia sclerotiorum*	*Fusarium* spp.	*Penicillium oxalicum*
Characteristic symptoms	Grey mould (conidia & conidiophores)	Brown/black specks (perithecia & pycnidia)	Dense white felt, black bodies (sclerotia)	Pink mould (sporodochia consisting of conidiophores & conidia)	Blue/green mould (conidiophores & conidia)
Symptoms on other parts of plant:					
● leaves	+	+	−	−	−
● collar and/or stem base	−	+	−	+	+
● fruit	+	+	+	−	+
	(165–362)	(163, 247, 363)	(365)	(249)	(248)

CUCUMBER

275 Reddish-brown/brown decay in a portion of stem covered in places with initially white masses turning brown (sclerotia)—*Sclerotinia sclerotiorum.*

275

276 Fibrous decay, beige/brown, initially wet, covered by salmon-pink mould—*Fusarium* sp.

276

277

277 Stub of rotted petiole covered with grey mould which provides nutrition for the parasitic fungus which will later invade the stem—*Botrytis cinerea*.

278

278 Shrivelled remains of petioles bearing numerous tiny black specks (pycnidia, perithecia); the fungus has invaded the stem which is beginning to rot—*Didymella bryoniae*.

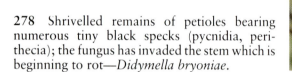

COURGETTE

The courgette and gourd, both mainly cultivated in the open, are far less liable to be affected by the fungi studied in this section than are melons and cucumbers. When conditions under cover are very humid, *Botrytis cinerea* will invade not only senescent tissues or wounds (**277**) but will cause damage, rarely serious, to the stem and leaves also (**167**); this is not the case with the fruit (**367–368**). *Didymella bryoniae* is rarely found on courgettes. It is able to invade senscent tissue (petioles, etc.) but does not continue to develop (**278**). It does cause damage to the collar and fruit (**369, 346–347**).

Internal decay of the stem

POSSIBLE CAUSES

- *Erwinia carotovora* **var.** *carotovora* (see page 125)
- *Erwinia tracheiphila* (factfile 3)

- *Fusarium oxysporum* **f.sp.** *cucumerinum* (factfile 22)
- *Fusarium oxysporum* **f.sp.** *melonis* (factfile 23)
- *Fusarium oxysporum* **f.sp.** *niveum* (factfile 24)
- *Verticillium dahliae* (factfile 25)

ADDITIONAL DIAGNOSTIC GUIDELINES

Cucurbits are affected by several vascular micro-organisms which invade and continue to penetrate the plant vessels. Once these are colonised, they generally become yellow and brown to a greater or lesser extent (**280**) and later the adjacent tissue also decays (**281**). That is the point at which one or several of the following symptoms become visible on the outer stem:

- Yellowing and browning of one side of the stem, presence of gummy globules (**282**);
- Flattening and longitudinal necrosis on one side of the stem (**283**);
- Whitish mould on the surface of the decayed tissues (**284**).

Very often cucurbits react strongly to vascular pathogens. To hinder their invasion of the vessels, the plants produce sticky substances, often in large quantities, which may be excreted to the outside of the stem in the form of dark brown globules (**282**). They are quite distinctive and give the name 'gummosis' to these vascular diseases.

The vascular location of these micro-organisms and the plant's reactions cause characteristic **wilting, yellowing and desiccation** (see also page 105).

To identify the vascular diseases the plants must be examined very carefully for the symptoms described in the following pages.

You will find that these diseases, in particular the specialised forms of *Fusarium oxysporum*, cause very similar symptoms in the various cucurbits which they affect.

279 Interior of normal stem—P = pith, V = vessels, B = bark (epidermis + cortical tissue).

280 Slight vascular yellowing and browning—*Verticillium dahliae.*

281 Marked vascular browning, necrosis in adjacent tissues—*Fusarium oxysporum* f.sp. *cucumerinum.*

282 Unilateral browning of the stem accompanied by exuded gummy globules which oxidise rapidly and become brown—*Fusarium oxysporum* f.sp. *melonis.*

283 The stem becomes necrotic and desiccates, becoming flat on one side—*Fusarium oxysporum* f.sp. *melonis.*

284 The fungus may fructify on the decayed tissues, forming a whitish-pink mould—*Fusarium oxysporum* f.sp. *cucumerinum.*

	MELON		CUCUMBER		WATER MELON
	Fusarium oxysporum f.sp. *melonis*	*Verticillium dahliae*	*Fusarium oxysporum f.sp. cucumerinum*	*Verticillium dahliae*	*Fusarium oxysporum f.sp. niveum*
Type of crops affected:					
• field grown	+	+	−	−	+
• protected	+	+/−	−/+	+	−
Symptoms in stem:					
• vessels yellow to light brown	−	+	−	+	−
• vessels dark brown	+	−	+	−	+
• decay of tissues close to vessels	+	−	+	−	+
Presence of gummy exudates and stem decay	+	−	+	−	+
Recovery of diseased plants	−/+	+	−/+	+	−/+

285

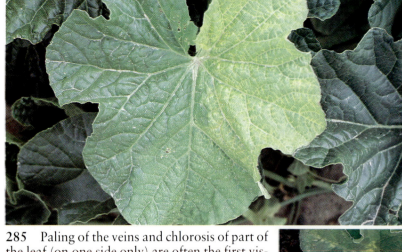

285 Paling of the veins and chlorosis of part of the leaf (on one side only) are often the first visible symptoms—*Fusarium oxysporum* f.sp. *melonis*.

287

286 The larger part of the leaf rapidly yellows, then wilts and becomes desiccated—*Fusarium oxysporum* f.sp. *melonis*.

288

287 Gummy globules, often brown, appear on the stem, preceding a unilateral beige/reddish-brown decay—*Fusarium oxysporum* f.sp. *melonis*.

289

288 Initially the vessels are often distinctly brown—*Fusarium oxysporum* f.sp. *melonis*.

289 Later, the pith and the cortical tissues become brown and spongy—*Fusarium oxysporum* f.sp. *melonis*.

290

291

290 Yellowing and desiccation of several leaves, often the oldest—*Verticillium dahliae.*

292

291 Unilateral wilting of the lamina accompanied by slight chlorosis—*Verticillium dahliae.*

293

292 Later the lamina shows inter-veinal necrosis and desiccation—*Verticillium dahliae.*

293 The vessels have slight yellow discoloration; no gummy exudate has appeared—*Verticillium dahliae.*

Very frequently, where there is fusarium wilt, a highly characteristic smell of honeysuckle accompanies the symptom of yellowing leaves.

296

294 Yellowing of old leaves—*Fusarium oxysporum* f.sp. *cucumerinum*.

296 A longitudinal area of necrosis can be seen on the stem (as on the melon) which is sometimes covered by gummy exudates and a whitish mould—*Fusarium oxysporum* f.sp. *cucumerinum*.

295 Yellowing on the leaf is on one side only; gradually the lamina wilts and becomes desiccated—*Fusarium oxysporum* f.sp. *cucumerinum*.

297 Marked browning of the vessels—*Fusarium oxysporum* f.sp. *cucumerinum*.

CUCUMBER

Fusarium wilt

298 V-shaped chlorotic lesions on the lamina; later the tissues become necrotic and desiccated—*Verticillium dahliae.*

299 Unilateral yellowing can be seen also on the petiole—*Verticillium dahliae.*

300 Unlike the effect of fusarium wilt, the vessels are only slightly discoloured, appearing dark yellow—*Verticillium dahliae.*

Verticillium wilt

CUCUMBER

301 Rapid wilting and desiccation of leaves on one or several branches which sometimes have a tendency to turn brown at the tips—*Fusarium oxysporum* f.sp. *niveum*.

302 Gummy exudates can be seen on the browned and desiccated stems—*Fusarium oxysporum* f.sp. *niveum*.

303 The vessels in the stem are dark yellow/brown—*Fusarium oxysporum* f.sp. *niveum*.

WATER MELON

Fusarium oxysporum f.sp. *melonis*

Today there are numerous varieties of melon which are resistant to certain strains of *Fusarium* (see Appendix III). This must be taken into account when making your diagnosis, but take care; there are several strains of the fungus (0, 1, 1–2). The disease causes fairly distinctive symptoms which enables it to be easily identified. Remember that some forms of the Race 1–2 cause rapid wilting (hence 'Wilt' disease) without preliminary yellowing or stem necrosis; in such cases, identification is more difficult. If you have any doubts at all, consult a specialist laboratory where the organism responsible can be isolated.

● *Verticillium dahliae*

This soil fungus is found fairly frequently in spring in melon-growing areas; it causes yellowing and foliar necrosis. However, we have come across it only in wet, cold springs which encourage root decay. Although very rarely found in such regions on melon plants, the organism has taken advantage of adverse growing conditions to become established. As soon as conditions become more favourable to growth, the plants make a rapid recovery. Recently, in several North European countries, various reversible symptoms have been noted on cucumber plants grown under protection. The vascular browning caused by this fungus in melon and cucumber plants is much less marked than the discoloration produced by the specialised strains of *Fusarium oxysporum*. If you are in any doubt, consult a specialist laboratory.

● **Fusarium oxysporum f.sp. *niveum***

Fusarium oxysporum f.sp. *benincasae* has been blamed for typical vascular symptoms on melon plants grafted on *Benincasa cerifera*. This *Fusarium* also invades several cultivars of the water melon, but not melon, and appears to differ from the other specialised forms dependent on melon, water melon and cucumber plants.

We should also mention *Fusarium oxysporum* f.sp. *lagenariae*, a vascular parasite fungus of *Lagenaria* sp. used as rootstock for the water melon, to control *Fusarium oxysporum* f.sp. *niveum*. In Taiwan there is also *Fusarium oxysporum* f.sp. *momordicae* found on *Momordica charantia* (bitter cucumber).

Quite recently the **parasitic specificity** of the vascular parasitic *Fusarium* affecting the Cucurbitaceae has been questioned. Dutch workers have shown that these specialised forms of *Fusarium* were able, under their experimental conditions, to infect several members of the Cucurbitaceae. There is no reason to believe that these results would be repeated in the field.

● *Erwinia tracheiphila*

This bacterium is not found in Europe but in the USA it causes considerable damage, particularly in melon and cucumber crops, and to a lesser degree in other species of *Cucurbita*. The water melon is virtually immune. We must emphasise here that this non-pectolytic *Erwinia* propagates in the vessels of certain Cucurbitaceae. The first signs are the wilting of several leaves, then whole branches. If a transverse cut is made through the stem, the latter will be found to contain a gluey substance, very characteristic of the disease, which forms a thread if the two parts of the stem are separated very gently.

Abnormalities and decay in flowers and fruit

SYMPTOMS EXAMINED

- Abnormalities and decay occurring during or just after the flowering period
- Small oily spots, necrotic and/or corky, on fruit
- Spots of various extent on fruit
- Decay affecting one side of the fruit
- Decay at the stylar or peduncular end of the fruit
- Decay in fruit resulting in a wet rot
- Cracks, splits, etc. of various sizes in fruit
- Fruit discoloration
- Changes in shape of fruit.

POSSIBLE CAUSES

- *Pseudomonas syringae* pv. *lachrymans*
- *Xanthomonas campestris* pv. *cucurbitae*

- *Botrytis cinerea*
- *Choanephora cucurbitae*
- *Cladosporium cucumerinum*
- *Cladosporium* spp.
- *Colletotrichum lagenarium*
- *Corynespora cassiicola*
- *Didymella bryoniae*
- *Diplodia natalensis*
- *Erysiphe cichoracearum*
- *Fusarium oxysporum* f.sp. *melonis*
- *Fusarium* spp.
- *Macrophomina phaseoli*
- *Mucor* spp.
- *Myrothecium roridum*
- *Penicillium digitatum*
- *Penicillium* sp.
- *Phytophthora capsici*
- *Pythiaceae* (*Pythium* spp., *Phytophthora* spp.)
- *Rhizoctonia solani*
- *Rhizopus nigricans*
- *Sclerotinia sclerotiorum*
- *Sclerotium rolfsii*
- *Sphaerotheca fuliginea*
- *Tricothecium roseum*

- Cucumber Mosaic Virus (CMV)

- Zucchini Yellow Mosaic Virus (ZYMV)
- Watermelon Mosaic Virus type 2 (WMV2)
- Cucumber Toad Skin Virus (CTSV)
- Papaya Ring Spot Virus (PRSV)
- Stylar distortion of cucumber fruit
- Appearance of male flowers in the all female cucumber varieties
- Blossom-end rot
- Apical flower bunches
- Nitrogen deficiency
- Sectorial chimera
- Corky stylar scarring
- Fruit drop
- Sun scald
- Check at primary stage in fruit formation
- Damage by hailstones
- Damage by birds
- 'Strangling' of the cucumber fruit
- Excess number of male flowers in the melon
- Splits during growth
- Bunched fruit
- Fissures in skin of cucumber fruits
- Oedema
- Yellowing of cucumber fruits
- Apical necrosis in water melons
- Fasciated nodes in the cucumber
- Various phytotoxicities
- 'Pillowy' fruit in the cucumber
- Giant sepals on cucumber flowers
- 'Glossy' appearance in the melon

OBSERVATION GUIDE

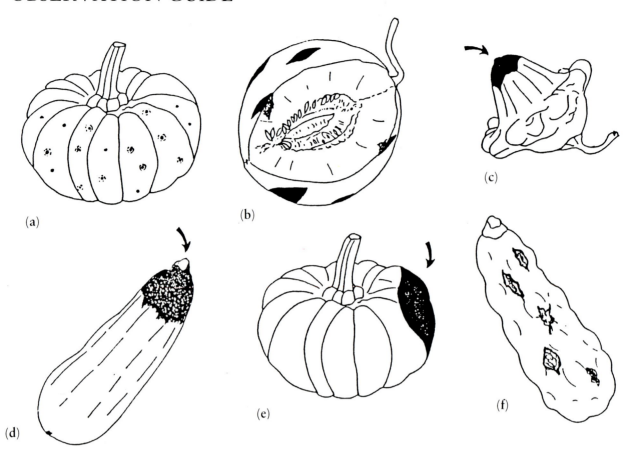

(a)

(b)

(c)

(d)

(e)

(f)

Location and appearance of decay in fruit

(a) Gourd with small spots, sometimes haloed.
(b) Fairly extensive spots on a melon, penetrating the fruit.
(c) Localised decay at the distal end of a pumpkin (stylar area).
(d) Localised decay at the apical end of a courgette (peduncular area).
(e) Localised decay on one side of a gourd.
(f) Dented and cracked courgette.

304

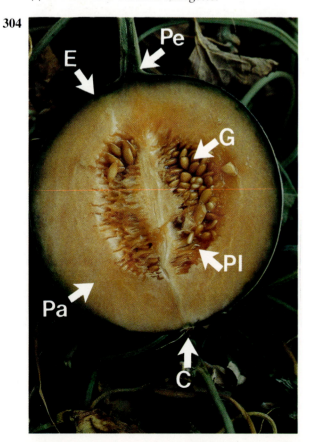

E : epidermis
Pe : peduncle (peduncular area)
Pa : parenchyma
Pl : placenta
G : seeds
C : stylar scar

Examples of deterioration in fruit

305 Patches of varying sizes on a melon fruit—*Cladosporium cucumerinum, Colletotrichum lagenarium.*

306 Cucumber fruit showing mottling—Cucumber Mosaic Virus (CMV).

307 Deformed and dented courgette—Zucchini Yellow Mosaic Virus (ZYMV).

308 Rotting tissue covered with grey mould developing at the end of a pumpkin fruit—*Botrytis cinerea.*

309 Clusters of male flowers spread over the main stem of a melon plant—Excess of male flowers.

310 Abnormal production of male flowers on a gynecious cucumber (P = petals, S = sepals)—Excess of male flowers.

311 Cucumber flowers with very large sepals—'Giant sepals'.

312 Cucumber plants carrying too many fruits at the inter-nodes—Bunched fruit.

Abnormalities and decay during or just after the flowering period

POSSIBLE CAUSES

- Excess of male flowers in the melon
- Appearance of male flowers in gynecious cucumber varieties
- Giant sepals on cucumber flowers
- Bunched fruits at nodes in cucumber plants
- Apical clusters of flowers
- Fruit drop

ADDITIONAL DIAGNOSTIC AIDS

- **Excess of male flowers in the melon**

In certain conditions which are not fully understood, numerous male flowers appear on melon plants, clustered at the nodes of the main stem (**309**). This transitory physiological phenomenon is not harmful to the plants and as a rule has no economic significance.

- **Appearance of male flowers in gynecious cucumber varieties**

Where these cucumber varieties are grown under glass, growers may sometimes find that a few plants bear large numbers of male flowers (**310**). This is a genetic anomaly and normally is not damaging to the crop as it occurs in only a very few genetically abnormal plants.

- **'Giant sepals'**

As in the case above, some aberrant plants may bear abnormal flowers with, for example, huge sepals (**311**). This can be considered as one of nature's curiosities.

- **'Bunched fruit at nodes'**

When there are sudden changes in the climatic conditions, the plants of certain cucumber varieties may produce an excessive number of female flowers (and therefore fruit) at the same level on the stem (**312**). This can be harmful to the plants and the subsequent fruit, which will remain small or even drop. For example, the cucumber variety 'Corona' is very prone to this abnormality, whereas the variety 'Girolla' is not susceptible. To avoid this defect, make sure that there are no sudden changes in environmental conditions.

- **Apical cluster of flowers**

Some viral infections (in particular ZYMV) cause growth retardation of the inter-nodes and a reduction in the size of the fruit. There may also be a cluster of flowers, sometimes more/less deformed, at the ends of the branches.

313

314

315

316

317

158

Fruit drop

The female flowers of the parthenocarpic varieties of cucumber do not need to be pollinated in order to set, and therefore produce, fruit. However, under certain conditions, for example when the mineral balance in the food supply is unsuitable (excess of nitrogen), or when temperatures are too high or too low, the developing ovaries become yellow and desiccated for no apparent reason (315), preventing fruit formation.

In monoecious varieties of cucumber and all varieties of melon, courgette, water melon and the other Cucurbitaceae, the female flowers must be pollinated to bear fruit. The lack of pollinators (bees, bumble bees) or male flowers in the courgette is likely to be responsible for drop, shown by the yellowing of the undeveloped fruit (314). Where a courgette has no pollinator and/or male flowers, it is possible to ensure parthenocarpic development of the fruit by the application of growth stimulators (procarpil, tomato-set).

Where melon and water melon are concerned, the plant's first priority is to supply the already well-developed fruit, preventing further fruit setting and stopping the growth of the young fruit (313). For this reason plants which set fruit very early and close to the stem base have slow vegetative growth and do not bear any further fruit. It is, therefore, advisable to remove the first fruits which have set too early.

As a general rule, plants which bear too much fruit and whose root system is damaged to a greater or lesser extent, may suffer from fruit drop.

'Airborne' parasitic micro-organisms

Several parasitic micro-organisms (fungi, bacteria, viruses) which cause severe outbreaks of disease before or just after the flowering period can be responsible for considerable blossom drop and the total destruction of the young fruit which becomes yellow and shrivelled. This happens, for example, in young cucumber fruits after infection by *Corynespora cassiicola*, or the causative agent of anthracnose, *Colletotrichum lagenarium* (317), and with certain viruses (ZYMV in melon and courgette) which also cause mottling of the petals (316). In milder outbreaks, the growth pattern of the developing fruit is seriously disturbed and this will adversely affect its final shape (401, 403).

---

313 Dull yellow melon fruit has lost its turgor and has stopped growing—Fruit drop.

315 Young cucumber fruit is becoming chlorotic and is wilting at the tip—Fruit drop.

314 Young courgette fruit is yellowing; the tip is shrivelling and becoming a reddish-brown/brown colour—Fruit drop.

316 Discrete mottling on the petals of a courgette flower—Zucchini Yellow Mosaic Virus (ZYMV).

317 Young cucumber almost completely rotted and oily—*Colletotrichum lagenarium*.

318 Small, oily, 'lip-shaped' spots on young melon fruit—*Cladosporium cucumerinum.*

319 Spots of brown canker, greasy on the periphery, on a courgette—*Cladosporium cucumerinum.*

320 Small, gummy cankers on a cucumber developing later into elliptical spots covered by a green felt—*Cladosporium cucumerinum.*

321 Small, dark green, wet spots on a cucumber fruit—*Pseudomonas syringae* pv. *lachrymans.*

322 Small, gummy, greasy cankers on cucumbers—*Colletotrichum lagenarium.*

Small oily spots, necrotic and/or corky fruit

POSSIBLE CAUSES

- *Pseudomonas syringae* pv. *lachrymans* (factfile 1)
- *Xanthomonas campestris* pv. *cucurbitae* (factfile 2)

- *Cladosporium cucumerinum* (**Gummosis**) (factfile 5)
- *Colletotrichum lagenarium* (**Anthracnose**) (factfile 6)

- Oedemas
- Various phytotoxicities

Refer also to the section: Cracks and splits of various sizes on fruit.

ADDITIONAL DIAGNOSTIC GUIDELINES

- *Cladosporium cucumerinum*

This fungus (responsible for **Gummosis**) can attack all the Cucurbitaceae; it affects mainly melon and courgette plants. Initially, small necrotic spots appear on the fruit (**318, 319, 320**) which may be overlooked on young fruit. The latter, especially courgettes, grow rapidly, and the spots which at first are innocuous can have very serious consequences at harvest time. The enlargement of the spots and the resulting formation of scar tissue cause considerable deformation of the fruit (refer to pages 168, 169).

In addition to the symptoms found on the fruit, other characteristic symptoms appear on the leaves and stems (see pages 74–76) which will enable you to confirm your diagnosis.

- *Colletotrichum lagenarium*

The frequent use of fungicides has almost eliminated this fungus responsible for anthracnose of cucurbits. Over the last decade we have observed only two outbreaks of melon and water melon plants, showing the rarity of the disease.

The symptoms it causes on fruit are fairly characteristic, especially in melons (**334–335**) and water melons (**336**). On cucumbers the symptoms are less typical. Growth often stops; there is yellowing and rotting in the very young fruit (**337**) or small greasy, gummy spots appear (**322**) which may be responsible for the fruit growing later into a crooked shape. Fruit which is already well-developed is far less susceptible. We advise you to refer also to the section on the damage this disease causes to leaves and stems, pages 74–76.

- *Pseudomonas syringae* pv. *lachrymans*

This bacterial disease, much feared by growers during the 1970s, has disappeared completely in France today. However, it is still present particularly in Eastern Europe and the USA, where it is reported to do great damage to water melon crops. Small, circular, brown spots, paler at the centre and with a yellow halo, appear on the fruit. In Italy, courgettes are very badly affected by brown spots, which vary in size and are surrounded by a yellow halo. The symptoms occur much more frequently on cucumbers: small, oily, circular spots (**321**) which sometimes rot when the bacteria have penetrated deeper, or after secondary infection by other bacteria. The exudate from the spots resembles teardrops, giving the bacterium its name. These symptoms may appear during storage.

323 Provençal musky gourd partially covered with small, beige, sticky spots, surrounded to a varying extent by a brown border—*Xanthomonas campestris* pv. *cucurbitae*.

324 Necrotic, sometimes sticky, spots with orange/yellow haloes on green fruit—*Xanthomonas campestris* pv. *cucurbitae*.

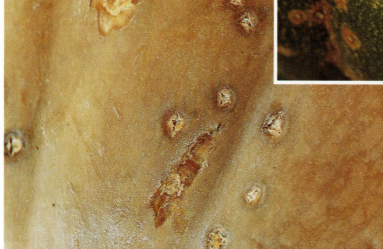

325 On ripe fruit the spots, as in the initial stages, are greasy and some are surrounded by an oily halo; the cuticle and the epidermis are splitting—*Xanthomonas campestris* pv. *cucurbitae*.

326 Brown rot, originating in a spot, penetrating deeper; in places the tissues are lysed to a varying extent—*Xanthomonas campestris* pv. *cucurbitae*.

GOURD

● *Xanthomonas campestris* pv. *cucurbitae*

This bacteria affects mainly gourd plants. Often it causes symptoms on leaves which are far from obvious but highly characteristic (**153–154**), but they appear initially on the fruit. As soon as the fruit ripens, especially in storage, the bacteria, located in the necrotic spots (**323, 324, 325**), penetrate into the flesh producing a reddish-brown rotted area (**326**), liquefying the flesh and infecting the seeds. It needs only a few rotted fruits in contact with healthy fruit to create foci of infection which will continue to spread.

   It has only recently been found in France. A factor which encourages its incidence is that some growers have adopted the bad habit of collecting some or all of their seed intended for future production from partially rotted fruit. Many of these seeds, and therefore seedlings, will be contaminated; the use of overhead irrigation or sprays ensures that the bacteria will be disseminated throughout the crop.

	Cladosporium cucumerinum	*Colletotrichum lagenarium*	Oedema	Various phytotoxicities
Very characteristic symptoms	Dark grey/green conidial felt	Salmon-pink gelatinous masses	No fructifications	No fructifications
Symptoms on other parts of plant	pages 75, 76, 78	pages 75, 76, 77, 80	page 101	pages 34, 62, 66, 68

327 Melon fruit covered with numerous oily spots—Oedema.

328 Later, these oily spots give way to tiny splits, more/less corky—Oedema.

329 Small patches of deterioration, more/less oily and raised—Phytotoxicity.

330 Numerous tiny.deterioration areas, oily and corky, on one side of a melon fruit—Phytotoxicity.

MELON

● Oedemas

Melons often show small, superficial damage which is very characteristic (327–328) and which growers tend to assume results from pests (puncture holes from thrips, etc.). But these symptoms are never caused by insects. It is high humidity, common under different forms of protection (tunnels, low continuous tunnels, etc) and especially beneath the vegetative canopy, which is to blame for this damage. Islets of water-saturated epidermal cells swell slightly and take on an oily appearance, while the distended cuticle finally splits (327, 328). The scar tissue which forms later resembles small corky pustules (331).

This non-parasitic disease is particularly common in spring, or when there is very high humidity affecting field-grown crops or crops under cover, and/or after over-generous watering, especially from overhead irrigation.

The damage is always superficial and affects only the appearance of the fruit.

● Various phytotoxicities

A few days after growers have sprayed pesticides on plants carrying young fruit (very susceptible at this stage), they sometimes notice that various blemishes have appeared on the fruit (329–330) which are fairly similar to those caused by oedema. In certain cases the leaves also are affected (see pages 34–62). The fragile cuticle of the young fruit, slightly damaged by the spray or sprays, is covered with numerous blemishes which are at first oily and more/less raised. These are often located on the most exposed face of the fruit. When spraying at high volume the suspension can accumulate under fruit, which is in contact with the plastic mulch. It is here that the damage will be found. Later, these blemishes form scar tissue, giving the cuticle a somewhat corky appearance (353).

331

331 After scar tissue has formed, small, beige, corky pustules can be seen on the surface of the melon—Oedema.

332 'Lip-shaped' brown necrotic spots with black centres—*Cladosporium cucumerinum.*

333 The depressed spots show a tendency to form scar tissue round the periphery; the centre is covered with felting, initially grey, turning dark green/black—*Cladosporium cucumerinum.*

334 Depressed patches, no scar tissue on the periphery, covered by salmon-pink fungal fructifications (acervuli sometimes forming concentric circles)—*Colletotrichum lagenarium.*

335 Round to oval patches, coloured brown/black; the centre may be covered by white mycelial masses as well as fungal fructifications—*Colletotrichum lagenarium.*

MELON

Spots of various sizes on fruit

POSSIBLE CAUSES

- *Botrytis cinerea* (factfile 7)
- *Cladosporium cucumerinum* (**Gummosis**) (factfile 5)
- *Colletotrichum lagenarium* (**Anthracnose**) (factfile 6)
- *Erysiphe cichoracearum* (**Mildew**) (factfile 9)
- *Sphaerotheca fuliginea* (**Mildew**) (factfile 9)

- **Damage by hailstones**
- **Pollen deposits**

ADDITIONAL DIAGNOSTIC GUIDELINES

- *Cladosporium cucumerinum, Colletotrichum lagenarium*

These two fungi are often mistakenly identified in the field as being the cause of certain foliar symptoms in the melon. While the symptoms may be confusing, these particular fungi should be perfectly easy to identify, if the photographs showing the specific symptoms they cause on fruit are studied carefully.

In the case of **gummosis**, the formation of corky scar tissue round the edges of the spots, particularly on the melon and courgette (**379–380**), give them a quite distinctive appearance. Almost all varieties are resistant to this disease; it is also very rare to find the small gummy spots (**320**) (sometimes covered by a grey-green down) which are characteristic of gummosis on cucumbers.

Anthracnose is now rarely found in crops of cucurbits. It is able to invade practically all the cultivated Cucurbitaceae, for example: melons (**334–335**), cucumbers (**322**), water melons (**336**) and even bottle gourds (**337**).

337 Dark green, greasy patches covered with pink gelatinous masses, on a water melon fruit—*Colletotrichum lagenarium*.

336 On a young bottle gourd, the spot may deteriorate rapidly into fruit rot—*Colletotrichum lagenarium*.

338   Young mildew colonies developing on a melon, as a white, powdery spot—*Sphaerotheca fuliginea*.

340   Grey, corky patches with very irregular outlines on a melon—Hailstone damage.

339   Very depressed, brown spots with distinct, darker edges, sometimes covered by grey mould, on a melon—*Botrytis cinerea*.

341   Numerous corky splits on a pumpkin fruit; the same type of deterioration can be seen on the stem—Hailstone damage.

342   Slightly powdery and raised small spots, ochre/fawn, on a melon; under the pollen, the tissues are translucent—Pollen deposits.

- *Erysiphe cichoracearum, Sphaerotheca fuliginea*

Erysiphe cichoracearum and *Sphaerotheca fuliginea* are the main mildews affecting the Cucurbitaceae. Normally they affect only the leaves and stems of certain cultivated cucurbits (**193–199**). In certain circumstances, especially when the inoculum is heavy and the climatic conditions favourable, more or less discrete powdery white spots may appear on the fruit (**338**).

- *Botrytis cinerea*

This fungus is a common cause of grey stylar rot on many cucurbits (**354, 362, 368**) and may also cause darkish, well-defined spots, very concave because of local rotting of the tissues (**339**). The way it infects the melon is quite specific: senescent tissues from the floral parts, especially the petals, already colonised by *Botrytis cinerea*, fall onto the young fruit. As these provide a reserve of food, infection takes place on contact, providing that the atmosphere is sufficiently humid, which it tends to be under the leaf canopy. Normally the number of melons affected is comparatively small.

- **Damage by hailstones**

The young fruit of melon, pumpkin and in general all the Cucurbitaceae are very fragile; hailstones can cause indelible blemishes at the point of impact (**340–341**). Other parts of the plant may also suffer, especially the peduncle and the stems (**266**). Normally, damage is widespread in the locality.

- **Pollen deposits**

Bees collecting nectar accumulate pollen in the form of a pellet, which may be quite large. If these happen to fall onto the leaves (**200**) and the fruit, especially melons and courgettes, they create little spots (**342**) which are usually quite harmless to the plant parts affected.

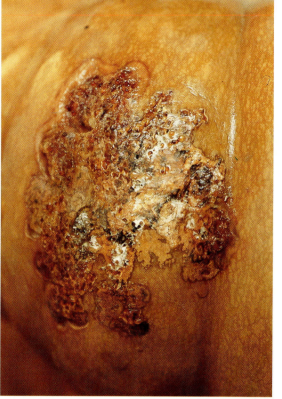

343. Young water melon fruit in which the part touching the ground has become dark green/black; the damaged tissues show a tendency to collapse, giving them a folded appearance, and are covered with a flat whitish felt—*Phytophthora capsici*.

345 Wet rot of part of a melon bruised by contact with the ground; black or blue mould appears in places—*Rhizopus nigricans*, *Penicillium* sp.

Side of fruit in contact with soil

344 Very irregular patch, depressed in places, with the surrounding fruit 'skin' often superficially fawn/brown. Small splits near the edge show discrete brown mycelial threads—*Rhizoctonia solani*.

346 Large, irregular, sticky and rotten area on the side of a courgette in contact with the ground; blacker areas can be seen and in places small, black spots (perithecia or pycnidia)—*Didymella bryoniae*.

Other fungi are able to invade the parts of fruit which are in contact with the soil. **Macrophomina phaseoli** causes wet, brown lesions covered with numerous black microsclerotia. **Sclerotium rolfsii** is responsible for rotting, which is rapidly invaded by a dense mycelial felt enclosing numerous small dark sclerotia, the size of a radish seed.

Deterioration affecting one side of the fruit

POSSIBLE CAUSES

- *Didymella bryoniae* (factfile 8)
- *Macrophomina phaseoli* (factfile 13)
- *Penicillium* sp.
- *Rhizoctonia solani* (factfile 15)
- *Rhizopus nigricans* (factfile 13)
- *Sclerotium rolfsii*
- Pythiaceae (*Pythium* spp., *Phytophthora* spp.) (factfile 14)

- Sun scald
- Various phytotoxicities (see also page 165)

ADDITIONAL DIAGNOSTIC GUIDELINES

The fruit of cucurbits are the most vulnerable part of the plant because of their fragile, watery tissues. Deterioration on one side of the fruit may be parasitic or non-parasitic in origin. There can be several explanations for this specific location of decay:

- If the damage is located on the side of the fruit in contact with the soil or the plastic mulch, it is frequently due to infection by soil-borne parasitic fungi. The part of the fruit touching very damp soil provides an ideal site for the penetration of several fungi. These develop either directly via the cuticle—a common path for various Pythiaceae (343), *Rhizoctonia solani* (344), *Didymella bryoniae* (346)—or through a variety of wounds. The latter is the usual path of opportunist saprophytic fungi of the genera *Rhizopus, Cladosporium*, etc. (345) (see also pages 183–184). Having colonised all the flesh, some of these fungi will also contaminate the seeds; this is particularly the case with *Didymella bryoniae* on gourds (347). Others behave in the same way in other parts of the plant; for example, *Rhizoctonia solani* and *Phytophthora capsici* which also invade the stem base;
- If the damage is located on the most exposed side, which is often the most easily visible, it is usually exposure to hot sun or phytotoxicities which is responsible for the more or less superficial burns on the fruit.

347

347 When the fruit is cut open it can be seen that the decay is spreading deeper into the flesh; the rotted tissue is invaded to a varying extent by the black mycelia of the fungus—*Didymella bryoniae*.

348 The placenta is quite rapidly colonised and the seeds are contaminated by the mycelium which invades the outer seed coats, colouring them brown/black—*Didymella bryoniae*.

348

349

350

351

352

353

172

- **Sun scald**

Cucurbit fruits are very delicate and their fragile watery tissues, when exposed to hot sunshine, may become dehydrated and overheated on that part which is facing the sun (**349, 350, 351**). Almost all cucurbits can be affected, especially when the foliage which normally protects them has been in some way damaged: e.g. invasion by airborne parasitic fungi and bacteria, which rots the leaves. All the colonised tissues may in turn be invaded by various secondary saprophytic fungi which eventually cause total decay (see page 183).

- **Various phytotoxicities**

Certain treatments by fungicides, etc., may also affect the fragile skins of cucurbit fruits. Some time after treatment growers may observe at the same time damage on leaves and fruit, although it is more often confined to the fruit. The superficial, corky blemishes (**353**) cover the part of the fruit which has been exposed to the spray and which has been slightly scorched as a result. The damaged tissues are stiff and eventually crack to a varying extent, when the fruits enlarge.

Side exposed to the sun . . .

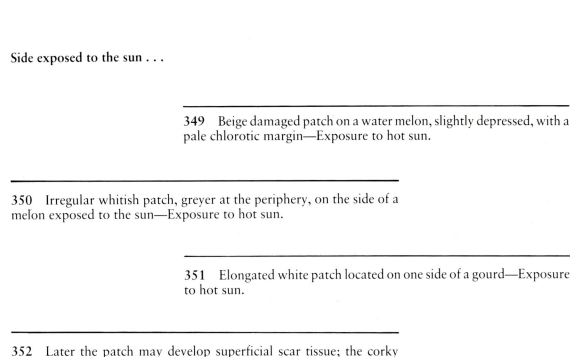

349 Beige damaged patch on a water melon, slightly depressed, with a pale chlorotic margin—Exposure to hot sun.

350 Irregular whitish patch, greyer at the periphery, on the side of a melon exposed to the sun—Exposure to hot sun.

351 Elongated white patch located on one side of a gourd—Exposure to hot sun.

352 Later the patch may develop superficial scar tissue; the corky tissues lose their elasticity and split as the fruit enlarges—Exposure to hot sun.

353 Melon fruit with the face exposed to spray treatments fairly uniformly cracked and corky—Various phytotoxicities.

354 Wet black rot covered by grey felt, located at the top of a melon—*Botrytis cinerea*.

355 Oily stylar rot partially covered by a thick white mycelium (fruit on right); later this forms large black sclerotia (fruit on left)—*Sclerotinia sclerotiorum*.

356 Splits around the peduncular area provide sites for the development of wet, sour rot—*Trichothecium roseum*.

357 Pink mould covering almost the whole rotted end section of a split melon—*Trichothecium roseum*.

358 Peduncular damage covered by pink mould on a melon; the vascular parasitic fungus has colonised the fruit after invading the plant vessels—*Fusarium oxysporum* f.sp. *melonis*.

Damage at the stylar or peduncular ends of the fruit

POSSIBLE CAUSES

- *Botrytis cinerea* (factfile 7)
- *Choanephora cucurbitae* (factfile 13)
- *Didymella bryoniae* (factfile 8)
- *Diplodia natalensis*
- *Fusarium oxysporum* f.sp. *melonis* (factfile 23)
- *Fusarium* spp. (see also page 183)
- *Penicillium digitatum*
- *Phytophthora capsici* (factfile 14)

- *Rhizopus nigricans* (factfile 13)
- *Sclerotinia sclerotiorum* (factfile 20)
- *Trichothecium roseum*

- Stylar abnormalities in cucumber fruit
- Corky stylar scarring
- Fruit drop (see page 157)
- Apical necrosis of water melon (Blossom end rot)

Why are these areas more susceptible?

Cucurbit fruits, like many others, possess natural entry sites which are liable to be colonised by specific fungi such as *Botrytis cinerea* (354), *Sclerotinia sclerotiorum* (355), *Trichothecium roseum* (356, 357), etc. Often these fungi can become established on a plant only if a source of nutrition is present (various senescent tissues such as desiccated flowers or petals, decaying leaves, etc.), or via a wound.

In cucurbit fruits these sites are most often found around the stylar scar. Withered petals tend to cling to this spot, and may do so for some time, if humidity is high. They are an ideal nutrient base upon which opportunist fungi can become established and later colonise the fruit; these are the organisms we shall be dealing with in this section. The stylar scar area is also very often the site of physiological splits (see page 188), open wounds that are extremely favourable for the penetration of these fungi as well as others whose efforts are described on page 182.

The peduncular scar is also vulnerable, though to a lesser extent. The peduncle may provide a nutrient base after the fruit has been picked or when the stem is damaged; for example: *Fusarium oxysporum* f.sp. *melonis* (358) or *Phytophthora capsici* (359). Natural splits which are a sign of maturity may occur on melons, and may be the site for colonisation by *Trichothecium roseum* (357).

359

359   Brown, wet rot, covered by a discrete white down developing on the face of a melon which was in contact with the ground—*Phytophthora capsici*.

	Botrytis cinerea	*Didymella bryoniae*	*Sclerotinia sclerotiorum*	*Trichothecium roseum*
Very characteristic symptoms	Grey mould visible to naked eye (conidiophores & conidia)	Tiny brown/black specks visible with a magnifying glass (perithecia or pycnidia)	Downy white mycelium; large white, later black, masses, hard to the touch (sclerotia)	Pink mould not to be confused with a *Fusarium* sp. (see page 182) (conidiophore and conidia)
Symptoms on other parts of the plant:	+	+	+	−
Melon		pages 74, 126, 138	page 138	
Cucumber	pages 92, 140	pages 92, 140	pages 140, 141	
Courgette	page 92 .			
Gourd		page 95		
Water Melon	page 94			

Other fungi can affect the stylar or peduncular areas of cucurbit fruits. Among those commonly found are *Diplodia natalensis* which causes decay beginning at the peduncle and a rapid black and wrinkled appearance to the fruit. *Penicillium digitatum* forms a blue mould on the apical tip of melon fruits.

ADDITIONAL DIAGNOSTIC GUIDELINES

- **Various Pythiaceae (*Phytophthora capsici*)**

Several *Phythium* (*P. aphanidermatum, P. ultimum, P. debaryanum*, etc.) and *Phytophthora* (*P. capsici, P. nicotianae*, etc.) can invade the soft tissue of cucurbits, such as the apex, the young shoots and also the young fruit. In France, it is **Phytophthora capsici** which causes the most damage, affecting melons, courgettes (370–372) and gourds (371). The damage caused is often similar: it looks wet, the tissues become discolored and sometimes shrivelled. On the surface there is a white felt, rather flat and sparse. The attack may take place in three places:
- At the stylar extremity (370), where the highly susceptible petals apparently provide an excellent nutritional base;
- At the peduncle (371), often following direct colonisation of this area, or of the adjacent stems;
- Directly in contact with wet soil (343).

- **Large corky stylar scar**

This is a non-parasitic disease, occurring infrequently. It has little effect on monoecious varieties of Charentais melon in which the stylar scar is generally quite small. In the andromonoecious varieties, where the scar is more marked, the disease is very harmful (360). Outbreaks occur mainly on early crops (under protection and field grown) in unfavourable conditions, in particular when temperatures fall too low during the flowering and fruit setting periods.

- **Stylar abnormalities in cucumbers**

Some varieties of cucumber produce fruit whose stylar end tip is described as a 'courgette end' (361). This is a genetic distortion and is therefore linked to the variety. Elimination of this trait during selection should enable growers to avoid the problem of finding such distortions in their fruit.

360 Exaggerated development of the corky stylar scar in a melon—Large corky stylar scar.

361

361   The cucumber on the left has a much more swollen stylar end than the fruit on the right, which is the normal shape for a cucumber—Stylar abnormality in cucumbers.

362

363

365

364

366

178

● *Choanephora cucurbitarum, Rhizopus nigricans (Rhizopus stolonifer)*

These two fungi are omnivorous saprophytes, which can cause rotting of the fruit of various cucurbits.

At present there are no reports of **Choanephora cucurbitarum** in France or UK, but it is found in tropical regions. It is a passive invader following colonisation via withered flowers (petals and sepals) or wounds. It causes a wet stylar-end rot which rapidly becomes covered with an abundant black aerial mould. Although omnivorous, this fungus is found most frequently on gourds, but also sometimes on cucumbers.

Rhizopus nigricans is quite commonly found, causing fruit rot. It produces symptoms which are very similar to *Choanephora cucurbitarum*, but less frequently invades the stylar-end of the fruit. It often colonises cucurbits (particularly melons and courgettes) via various wounds, producing a black mould on the decayed tissues which is very similar to that caused by *Choanephora cucurbitarum* (**374**).

362 Cucumber with rotted end; all the deteriorated tissues are covered by grey mould; the dark green area of the advancing edge contrasts with the still healthy portion of the fruit—*Botrytis cinerea.*

364 The black colour of the rot comes from the very numerous fungal fructifications in the damaged tissues—*Didymella bryoniae.*

363 Brown deterioration at the cucumber ends, which also appear constricted; the adjacent tissues are often pale green/yellow—*Didymella bryoniae.*

365 White 'spiral' mycelia encircling a cucumber three-quarters of which is rotten—*Sclerotinia sclerotiorum.*

366 This internal rot of the fruit will make it unmarketable—*Didymella bryoniae.*

CUCUMBER

367

368

367 Young courgette fruit with the pathogen spreading rapidly to the fruit from its infected and rotting petals—*Botrytis cinerea*.

369

368 Rot has developed at the end of this courgette which is becoming soft and wet—*Botrytis cinerea*.

370

369 As in the cucumber, this courgette shows a dark reddish-brown rotted area and the end of the fruit is constricted—*Didymella bryoniae*.

370 Browning of the petals and the ends of a young courgette; a dark exudate can be seen on the affected tissues which have a diffuse border—*Phytophthora capsici*.

COURGETTE

● *Phytophthora capsici*

This relatively polyphagic pathogen does not exclusively infect peppers. Cucurbits are quite susceptible, in particular, species of *Cucurbita* and to a lesser extent, the melon. It can cause decay of the plant stems and collars (**231–262**), and also of the fruit (**370, 371,** etc.). The damaged tissues look wet and are often covered by a rather flat white mycelium when weather conditions are especially humid or after applying water.

Various **Pythium** species, particularly **Pythium aphanidermatum** in tropical areas, gain entry to the fruit via wounds or by using the floral parts to establish. They initiate a dark green rot, which is rapidly covered by a downy white mycelium.

● **Water melon apical necrosis (Blossom-end rot)**

Large brown decayed areas sometimes appear at the ends of water melon fruits. Affected tissues wilt; as the fruit grows the central tissues gradually decompose to an increasing amount. This physiological disorder is supposed to be caused by too little, or no, water (following badly managed irrigation, root losses caused by pathogens or cultural operations) and may be linked to calcium deficiency. *Pythium aphanidermatum* and *Pythium debaryanum* may possibly be associated with this symptom.

371

371 Clearly defined brown decay, starting in the peduncle of a gourd—*Phytophthora capsici.*

372

372 End of a rotted courgette; note the presence of a white growth consisting of bacteria mixed with very numerous fungal sporangia—*Phytophthora capsici*.

373

373 A melon which has burst open at the stylar scar; the exposed flesh provides ideal nutrition for the pink mould, which has partially colonised it—*Fusarium* sp.

374

374 Young cucumber partially covered by a dense mould with numerous black pin-heads—*Rhizopus nigricans*.

375

375 Ripe gourd covered by numerous velvety, dark green/black colonies—*Alternaria* spp.

Wet, deliquescent rot in fruit

POSSIBLE CAUSES

- *Erwinia* spp. (*Erwinia carotovora*)
- *Xanthomonas campestris* pv. *cucurbitae* (factfile 2)
- *Xanthomonas campestris* pv. *melonis*
- *Alternaria tenuis*
- *Botrytis cinerea* (see pages 174, 178, 180) (factfile 7)
- *Choanephora cucurbitarum* (factfile 13)
- *Cladosporium* spp.
- *Didymella bryoniae* (see pages 178–180) (factfile 8)
- *Fusarium* spp.
- *Geotrichum candidum* (factfile 13)

- *Mucor* spp.
- *Penicillium* spp.
- Pythiaceae (*Pythium* spp., *Phytophthora* spp.) (factfile 14)
- *Rhizopus nigricans* (factfile 13)
- *Sclerotinia sclerotiorum* (see pages 174–178) (factfile 20)
- *Trichothecium roseum*

- 'Pillowy' cucumber fruit
- 'Glassy' areas in melon fruit

Refer also in particular to the preceding section.

ADDITIONAL DIAGNOSTIC GUIDELINES

Numerous micro-organisms are responsible, individually or in combination, for fruit rots in cucurbits. As a rule, the deterioration is soft and wet and the flesh quickly collapses. Among them are species of *Erwinia* (*Erwinia carotovora*, *Erwinia aroideae*, *Erwinia ananas*, *Erwinia carnegieana*, etc.) and *Xanthomonas* (*Xanthomonas campestris* pv. *cucurbitae* and also *Xanthomonas campestris* pv. *melonis*, which is common in Brazil but differs from the preceding species). There are also many fungi such as *Botrytis cinerea*, several species of *Fusarium* (*Fusarium roseum*, *Fusarium solani* f.sp. *cucurbitae*, race 2); Mucoraceous fungi (*Mucor* spp., *Rhizopus nigricans*, *Choanephora cucurbitae* etc.); various Pythiaceae (*Pythium aphanidermatum*, *Pythium ultimum*, *Phytophthora capsici*, etc.).

All these micro-organisms have certain characteristics in common:

- They penetrate the fruit via various lesions (e.g. insect punctures, growth splits, perforations, microscopic lesions of the cuticle) or by first establishing on senescent tissue (desiccated petals and sepals, etc.);
- They are greatly influenced by the agricultural and climatic conditions during cultivation. For example, the very high humidity which results from heavy rainfall or irrigation encourages their development;
- They invade unripe fruit but are more frequent on mature or even overripe fruit. With a few exceptions, species which are not supported so that the fruit is on the soil are the most vulnerable to damage;
- They cause very similar types of rot, which develops rapidly and is accompanied by wetness and collapse of the internal tissues. They only feature by which a causal organism can sometimes be distinguished is by the appearance of the fungal fructifications or the bacterial ooze on the surface of the fruit (refer to the table on the following page).

All these micro-organisms are active on their own or in association with others; several cause considerable damage during transport and storage of the fruit.

376

376 Badly developed, senescent cucumber covered by numerous blue and black moulds, the source of the rot—*Alternaria* sp., *Penicillium* sp., *Cladosporium* sp.

Symptoms	Micro-organisms	Commonest susceptible hosts
Creamy mould		
● whitish	*Erwinia* spp.	all cucurbits
	Geotrichum candidum	all cucurbits
● yellowish	*Xanthomonas* spp.	
	X. cucurbitae	Gourd
	X. melonis	Melon
Mycelial growth		
● whitish/grey	Various Pythiaceae,	Gourd
	esp. *P. capsici*	Courgette
		Melon
● grey	*Botrytis cinerea*	all cucurbits
● white/pinkish	*Fusarium* spp.	Melon
	(*F. solani, F. roseum*)	Gourd
● pink	*Trichothecium roseum*	Melon
● blue/green	*Penicillium* spp.	Melon
● dark green/black	*Cladosporium* spp.	Melon
	Alternaria tenuis	Melon
	Stemphylium botryosum	Melon
	Ulocladium botrytis	Melon
Dense mycelium with numerous black pinheads	*Rhizopus nigricans*	Melon, courgette
	Choanephora cucurbitarum	Gourd
Dense white growth covered by black structures	*Sclerotinia sclerotiorum*	Melon, cucumber

● **Pillowy fruit in cucumbers**

This non-parasitic condition of cucumbers occurs when they are stored at low temperatures soon after harvesting. Opaque, white, porous areas can be seen inside the fruit, especially in the mesocarp. The parenchymatous cells are hypertrophied and very numerous. The cause of the condition appears to be connected with calcium deficiency.

● 'Glassy' areas in melon fruit

This is a non-parasitic disease of melon that is not yet fully understood. Its trademark is the 'glassy' appearance (377) of the flesh in the area between the rind and the seed cavity. The flesh rapidly becomes deliquescent and gives off a smell of alcoholic fermentation. The entire fruit is affected; at a very advanced stage the surface appears wet (378). Among the factors which favour this disease are the following:

● Adverse conditions affecting soil and climate; non-vigorous plants, planted out into cold soils providing a poor environment for root development, will succumb more easily. The plants will also be more susceptible after periods of overcast weather in soils starved of oxygen, when there are considerable differences between day and night temperatures, etc.;
● Unsuitable mineral balance, with deficiencies or excesses;
● Calcium deficiency and over-watering while the fruit is swelling and ripening;
● Varietal differences in susceptibility occur but have not yet been clearly established.

Because of its highly characteristic symptoms the disease is easier to identify than it is to explain. By adopting the following measures you may be able to avoid it:

● Plant out only when conditions are suitable, and never into soil which is too wet or too cold;
● If the irrigation water has a low calcium content (less than 150 mg/l) calcium should be added as soon as the male flowers appear;
● Adjust watering according to growth stage and weather conditions: ensure a plentiful supply as the fruits begin to ripen, but reduce the amount a week before harvesting, and when the weather is overcast.

377 Melon fruit cut open through the centre showing 'glassy' flesh—'Glassy' areas in melon fruit.

378 Irregular oily patches visible on the outside of the fruit—'Glassy' areas in melon fruit.

379 Melon with numerous longitudinal, corky cracks of various sizes; some have a dark green mould on the surface—*Cladosporium cucumerinum*.

380 Courgette fruit deformed by numerous corky cracks—*Cladosporium cucumerinum*.

381 Transverse and longitudinal corky cracks covering a melon—Zucchini Yellow Mosaic Virus (ZYMV).

382 The flesh of the fruit is marbled and has an abnormally firm consistency—Zucchini Yellow Mosaic Virus (ZYMV).

- **Zucchini Yellow Mosaic Virus (ZYMV)**

This virus causes two types of symptoms, which sometimes occur together, on melons:

- Superficial cracks, which may be deeper, with corky edges (**381**);
- Marbling of the flesh (**382**) which becomes harder making the fruit inedible and therefore unmarketable. This symptom appears more frequently in combination with Cucumber Mosaic Virus (CMV).

Cracks and splits of various sizes (sometimes corky) on fruit

POSSIBLE CAUSES

- *Xanthomonas campestris* pv. cucurbitae (factfile 2)
- *Cladosporium cucumerinum* (factfile 5)
- *Myrothecium roridum*
- Zucchini Yellow Mosaic Virus (ZYMV) (factfile 32)
- Bird damage
- Growth splits
- Epidermal splits in cucumbers
- Various phytotoxicities (see pages 164–173)

ADDITIONAL DIAGNOSTIC GUIDELINES

Young cucurbit fruits have a very fragile cuticle; a harmless lesion at this stage may render the fruit unmarketable when harvested (383). Damage from various sources, and sometimes small lesions caused by parasitic micro-organisms on the skin, may develop corky scar tissue. This tissue is not elastic and when the fruit swells the skin cracks creating splits of various sizes. Thus what may start as a harmless blemish on a young fruit becomes more extensive as the fruit grows, with sometimes spectacular results (379–380). The earlier that damage occurs on the young fruit, the greater will be the final damage. There are many causes of fruit cracks and splits; this section will summarise the most common.

- *Cladosporium cucumerinum*

The effects of early outbreaks on young melons or courgettes can severely reduce the commercial value of the fruit (see 379–380). Usually, these symptoms are preceded, or accompanied by, symptoms on the leaves and stems (highly visible on melon plants, not so obvious on courgettes) which enables the disease to be positively identified (refer to pages 74, 76, 78).

- *Myrothecium roridum*

Reported in southern Texas, in the USA, this fungus causes superficial, deep or hollowed areas of damage ranging in size from 2–50 mm diameter and covered with dark green/black fungal sporolation (sporodochia). It also causes leaf spots (see page 89).

383

- Various lesions

Figure 383 shows a melon deformed by corky cracks. It is a good illustration of how lesions, apparently harmless when they occur on very young melons (mechanical damage in this case), can result in unmarketable fruit.

384

385

386

387

388

● *Xanthomonas campestris* pv. *cucurbitae*

This bacterium causes inconspicuous foliar symptoms (153–154); when infection is early in the development of the fruit small, prominent cracks result (384). This symptom occurs after the development of small, less obvious spots (323–324).

● 'Growth splits' in the melon

Every grower must have been, or one day will be, confronted with the problem of splits in melons (386) which result in the fruit bursting at the stylar end. The exact cause of this undoubtedly physiological condition is not known; it appears very likely that the fruit's nutritional supply, and especially water supply, are relevant factors. Plants which have been subjected to erratic watering patterns (wet periods succeeding dry) have a particular tendency to show this symptom. There are also wide differences in varietal susceptibility.

We would most strongly advise you to give up growing highly susceptible varieties, and to try to control your irrigation more carefully, especially when irrigation channels are used, so that less water is applied more frequently.

● 'Splits' in the skin of cucumbers

As we have seen, the skin of young cucurbit fruits is very tender and fragile. It can be injured by various agents such as the sun or pesticides, which cause superficial scorch. Cucumbers grown under cover sometimes develop very small longitudinal or transverse splits, light brown and corky, spread over one face of the fruit or the whole surface (385–387). These symptoms are due to temperature stress (cold draughts, a short period of low temperature, etc.) which affects the skin of young fruit grown under cover and exposed to cold air or to cold condensation dripping from the roof. The superficial lesions on the skin later become corky, ruining the appearance of the fruit for commercial purposes.

● Bird damage

Many birds are attracted to cucurbit fruit; pecking can cause considerable damage (388).

---

384   Gourd partially covered by small, corky, raised cracks, which may be sticky—*Xanthomonas campestris* pv. *cucurbitae*.

386   Melon with splits at the apical end giving numerous saprophytic fungi (which may be carried by fruit-flies [drosophilae]) access to the flesh—'Growth splits'.

388   Melon showing suberised open lesions, of various shapes, the result of birds pecking at the fruit—Damage by birds.

385   Small splits covering a large part of these cucumbers—'Splits in cucumber fruit skins'.

387   Later, the splits may become suberised and cause deformation of the fruit—'Splits in cucumber fruit skins'.

389 Discrete mottling on a melon fruit—Cucumber Mosaic Virus (CMV).

390 Obvious green mottling on a ripe fruit—Cucumber Mosaic Virus (CMV) + Water Melon Mosaic Virus, type 2 (WMV2).

391 Circular spots on unripe fruit—Zucchini Yellow Mosaic Virus (ZYMV).

392 Large, white, slightly raised patches on unripe fruit—Zucchini Yellow Mosaic Virus (ZYMV).

MELON

Fruit discoloration

POSSIBLE CAUSES

- **Cucumber Mosaic Virus (CMV)** (factfile 29)
- **Zucchini Yellow Mosaic Virus (ZYMV)** (factfile 32)
- **Water Melon Mosaic Virus, type 2 (WMV2)** (factfile 31)
- **Cucumber Toad Skin Virus (CTSV)** (factfile 34)
- **Various viruses**

- **Partial chimera**
- **Cucumber Yellows** (factfile 33)
- **Melon Yellows** (factfile 33)

ADDITIONAL DIAGNOSTIC GUIDELINES

- **Various viruses**

Most viruses are capable of causing external disfigurement in cucurbit fruits, perhaps making them unmarketable. The severity of the symptoms varies with the virus, the strain, the time the plant is infected, and the plant variety. In melon, mottling is a common symptom (**389–390**), but in other cases the symptom is more localised (**391–392**) and is combined with slight distortion (e.g. lumpiness). Mottling is often more pronounced on young unripe fruit and tends to become less distinct as the fruit ripens. It may also appear on the yellow varieties (of the Canary Melon type).

The symptoms are frequently insufficiently marked to justify a positive diagnosis and it should be noted that characteristic symptoms can often be seen on the leaves. Refer to the section 'Abnormalities in the Shape and Appearance of Fruit', page 195.

393

- **Cucumber Yellows**

Cucumber plants infected by Cucumber Pale Fruit Viroid may show uniformly pale green to yellow discoloured fruit. This viroid, thought to be a strain of Hop Stunt Viroid, has been reported in northern Europe.

- **Melon Yellows**

Growers may sometimes notice bright yellow areas on the sides of unripe smooth Charentais melons (**393**), spoiling the external appearance of the fruit. It is a local reaction of the rind when exposed to sunlight and tends to fade as the fruit ripens. Some varieties are more sensitive, especially monoecious varieties which tend to have sparse foliage.

393   Partially yellowed sides on a melon
Melon Yellows.

394 Pronounced mottling on fruit with a slightly lumpy surface—Cucumber Mosaic Virus (CMV).

395 Yellow mosaic on fruit—Zucchini Yellow Mosaic Virus (ZYMV).

396 Numerous necrotic and chlorotic dots on fruit—Cucumber Toad Skin Virus (CTSV).

397 Fruit partially yellowed or blanched—Chimera.

CUCUMBER

● **Partial chimera**

Parts of cucurbits above ground may show uniformly yellow or white sectors (leaves and stems, see pages 39–66), particularly on the fruit (**397**). This colour change is the result of a local mutation (a genetic anomaly), which affects chlorophyll production and therefore hinders photosynthesis. The condition is normally quite rare and does not affect the crop.

Virus	Melon	Cucumber	Courgette
CMV	Slight marbling (389)	Mottling and lumpy appearance (394)	'Dotted' appearance, slight deformity (398)
WMV2	Pronounced mottling	No symptoms	None, rarely a few light spots. Green mottling on yellow fruit varieties
ZYMV	Mottling causing slight deformation and cracks (381 and 391). Splitting.	Mottling and deformation (395)	Fruits very lumpy, sometimes discoloured (403)
PRSV	Marbling	Mottling	Lumpy fruit (405)
SqMV	Pronounced mottling in Charentais vars.	No symptoms	Considerable deformation
CTSV		Mottling, deformation, necrotic or yellow dots (396)	

399 Young mottled fruit with small lumps—Zucchini Yellow Mosaic Virus (ZYMV).

399

398 Discrete mottling on fruit, with 'dotted' appearance—Cucumber Mosaic Virus (CMV).

398

400

COURGETTE
AND
GOURD

400 Gourd with mosaic disease, showing lumpy areas and mottling—Zucchini Yellow Mosaic Virus (ZYMV).

193

401	Melon	402	Cucumber
403	Courgette	404	Gourd

As illustrated in **401, 402, 403** and **404**, fruit affected by Zucchini Yellow Mosaic Virus (ZYMV) is not only mottled but also may become lumpy and deformed.

Abnormalities in the shape and appearance of fruit

POSSIBLE CAUSES

- **Cucumber Mosaic Virus (CMV)** (factfile 29)
- **Zucchini Yellow Mosaic Virus (ZYMV)** (factfile 32)
- **Papaya Ring Spot Virus (PRSV)** (factfile 30)
- **Cucumber Toad Skin Virus (CTSV)** (factfile 34)

- **Fasciated fruit**
- **Constriction of cucumber fruits**
- **Various deficiencies**

ADDITIONAL DIAGNOSTIC GUIDELINES

- **Various viruses**

There are several viruses capable of causing more or less disastrous changes to the appearance of fruit. The most striking, and also the most damaging, is without doubt Zucchini Yellow Mosaic Virus. As far as the others are concerned, it is always advisable to be careful as the fruit symptoms are not enough by themselves to make a positive diagnosis. We would advise you to refer to the section 'Fruit Discoloration', and particularly to the table on page 193, in order to confirm your conclusions.

406 This fruit is slightly curved and is also partially grooved and slightly lumpy—Cucumber Toad Skin Virus (CTSV).

405 1. Fruit showing numerous small depressions (CMV).
2. Very deformed fruit, the result of the numerous lumps which may be quite prominent (ZYMV).
3. Fruit with more regular and less obtrusive lumps (PRSV).

407

408

409

410

196

- **Fasciated fruit**

Most cucurbit varieties may occasionally produce fruit which appears to result from the fusing of two fruits (Siamese fruit) (**407–408**). This is due to abnormal growth of the ovarian tissue which divides incompletely, thus giving two fused fruits. It is quite a rare condition and may be considered as a curiosity.

- **'Constriction' of cucumber fruit**

The fruit of the non-parthenocarpic varieties of cucumber is sometimes 'constricted' (**409**). The condition is the result of incomplete pollination: the ovules located in the area of least growth have not been fertilised and have therefore not developed into seeds.

Incomplete pollination of the fruit may be associated with a lack of male flowers or due to reduced activity of pollinating insects; this in turn is often linked to adverse weather conditions (e.g. low temperatures, rain). When plants have suffered root loss, the fruit may become constricted in one or more areas.

- **Various deficiencies in cucumbers**

Deformities in fruit may be caused by certain deficiencies. The most common are nitrogen deficiency, causing pinching of the fruit end (**410**), and potassium deficiency which produces constriction at the peduncular area.

Part Two

Main characteristics of the pathogenic agents and methods of control

BACTERIA

Factfile 1

Pseudomonas syringae pv. *lachrymans* (Smith & Bryan) Young, Dye & Wilkie.

Synonym: *Pseudomonas lachrymans* (Smith & Brian) Carsner.

Causes: **Angular spot disease**

SYMPTOMS
Refer to 148, 149, 150, 321.

MAIN CHARACTERISTICS

● **Frequency and economic importance:** This bacterium is found worldwide and affects numerous cucurbits, such as cucumbers and water melons in the USA, courgettes in Italy, etc. It was reported in the UK and France for many years but is now unusual.

● **Source:** Essentially this organism survives in plant debris which has been ploughed back into the soil. The organism can survive in that environment for up to 2 years and thus remains viable from one season to another. It may also live on the surface of seeds which were contaminated at the time they were collected, and subsequently act as sources. Another possible source is the irrigation water.

● **Spread:** The organism is disseminated by rain, irrigation water and wind, which carries water droplets. It may also be spread via clothing, tools or dirty hands during the course of cultivation work.

● **Penetration:** Via the stomata and a very wide diversity of wounds.

● **Conditions encouraging development:** As for numerous bacteria, ambient humidity is the main factor influencing the growth of this *Pseudomonas*. Its activity is greatly enhanced by the presence of free water on plants, particularly after heavy rain or overhead irrigation. Maximum infection takes place before dawn, because at that period the leaves are more or less covered with water because of the accumulation of dew. Two weeks of dry conditions are sufficient to stop the spread of the disease. The bacterium has an optimum temperature of between 24°C and 28°C; it is able to multiply at 36°C and so thrives in high temperatures. Excessive nitrogen dressings are thought to make plants more susceptible to this pathogen.

CONTROL

● **During cultivation:** As soon as the first symptoms are observed, apply copper in the form of Bordeaux mixture every 5–10 days, thoroughly wetting the vegetation. The association of copper with certain dithiocarbamates* such as maneb and mancozeb is likely to be more efficient and versatile, because it will also control many fungi affecting the foliage. Cucurbits are, however, fairly sensitive and these applications may cause phytotoxicities, if used in high concentrations in hot weather.

It is advisable to avoid overhead irrigation. Glasshouses should be well ventilated to reduce ambient humidity and prevent the possibility of free water on the plants' surface.

All debris must be cleared when the crop is harvested.

● **Subsequent crop:** Where crops are field grown it is advisable to rotate. This is not essential in tropical regions as the pathogen rarely survives in the soil.

Seeds must be clean and can be disinfected, for example, by heat.

Some varieties of commercially grown cucumbers in the USA are tolerant or resistant to this pathogen.

IMPORTANT NOTE
* Approval of fungicides may vary from country to country and local labels should be consulted before use.

(Bacteria)

Xanthomonas campestris pv. cucurbitae (Bryan) Dye.

Synonym: *Xanthomonas cucurbitae* (Brian) Dowson.

Causes: **Bacterial rot in gourd fruit**

SYMPTOMS

Refer to **153, 154, 323, 324, 325, 326, 384.**

MAIN CHARACTERISTICS

• **Frequency and economic importance:** Outbreaks of this bacterium have been reported in several countries, in particular Australia, India and the USA, on cucumbers and gourds. In France gourds in the south-east and the south-west of the country are the only crop affected. Damage to the fruit can be serious, especially during storage; 30% of the fruit may go rotten over the winter storage period.

• **Source:** The organisms survive mainly in seeds. Gourds are normally grown in the traditional manner, and very often growers will collect seeds from fruit from the previous season which are already contaminated. The bacteria can also survive in plant debris which has been buried in the soil.

• **Penetration:** Generally through the stomata of the leaves. On the fruit, we have come to the conclusion that the bacteria gain access via the lenticels or through insect punctures.

• **Spread:** Like many other bacteria they are spread via rain, spraying and wind, which transports water droplets over considerable distances.

• **Conditions encouraging development:** Not a great deal is known about the biology of this bacterium. It appears during the summer, when temperatures are fairly high, and frequently after heavy rain or overhead irrigation.

CONTROL

• **During cultivation:** There is no effective control for this disease. We advise you to ensure that humidity is controlled, and especially that free-standing water does not collect on the plants. Avoid overhead irrigation as it encourages the growth and dissemination of the bacterium. If you have to irrigate overhead, do so in the morning (never at night), so that the foliage can dry out quickly during the day.

Spraying with copper in the form of Bordeaux mixture (ensuring that the plants are thoroughly wetted) will check the disease, particularly if the treatment is applied early and repeated after rain or overhead watering.

During and at the end of cultivation destroy all plant debris, in particular the affected fruit. Only pick fruit which appears to be sound.

• **Following crop:** It is of the utmost importance that the seeds should be healthy, gathered from healthy fruit. Select the latter with the greatest care. Never collect seeds from affected fruit—it is false economy. Seeds can be disinfected in sodium hypochlorite, although it is not certain that this treatment is effective.

Where the soil has become infested, a plan of preventive treatment, using copper-based products, should be started at the seedling stage, and continued throughout the growing season.

The measures suggested for the control of *Pseudomonas syringae* pv. *lachrymans* can also be used.

Erwinia tracheiphila (EFSM) Halland

Causes: **Bacterial wilt**

SYMPTOMS

Refer to **151**.

MAIN CHARACTERISTICS

- **Frequency and economic importance:** This bacterium is found in several countries in Asia, Africa and especially North America. It affects several cucurbits, both cultivated and wild, but melon and cucumber are the crops which suffer the most damage. It has been reported in Europe, but not at present in the UK or France, perhaps due to the absence of insect vectors.

- **Source:** During the winter it survives either in the digestive tracts of various plant feeding bottle coleopteran vectors or in affected hosts. It can live for 1–3 months in the buried part of stems, but apparently cannot survive from one season to the next in this way.

- **Spread:** The organism is disseminated by plant-feeding phytophagic coleopterans with cucurbit hosts, e.g. *Acalymma vittata* and *Diabrotica undecimpunctata*. These acquire the bacterium by feeding on leaves contaminated by bacterial ooze which they then inoculate into other plants during the process of feeding.

Unlike many *Erwinia* species it is not pectolytic (does not cause rotting) but rapidly penetrates into the vessels of the stem and then moves downwards, in contrast to other vascular diseases. It is not transmitted in the seeds.

- **Conditions encouraging development:** Environmental conditions such as temperature or humidity do not appear to influence the growth of this bacterium; but it is most active during a dry period or after heavy rainfall. The disease is most likely to occur under these conditions.

CONTROL

- **During cultivation:** When an outbreak is identified in the growing crop, the first plant affected must be destroyed at once and insecticides applied as soon as possible, to eliminate the insect vectors. The spraying of copper compounds has no effect on this disease.

- **Following crop:** To be effective, chemical sprays to eliminate the insect vectors should be applied early as a preventative measure. 'Decoy' plants, treated with insecticides, may be used to reduce the insect population.

All the commercial varieties of melon and cucumber at present cultivated in France are susceptible to this bacterium. In the USA, some cucumber varieties with the *Bw* gene are resistant.

Other bacteria infecting the Cucurbitaceae

Several species of *Erwinia* (**Erwinia ananas, Erwinia aroidaeae, Erwinia carnegiana**), and in particular **Erwinia carotovora** var. *carotovora*, can cause damage to the stems (245) and fruit of various cucurbits due to the secretion of numerous cellulolytic and pectolytic enzymes. These bacteria are very polyphagous and become active when humidity is high and temperatures vary between 5°–37°C, with an optimum of 22°C. They survive in soil on the debris of diseased plants.

Outbreaks normally occur in crops grown under protection, less often in the field, and the control methods used are the same as those for *Pseudomonas syringae* pv. *lachrymans*.

When the collar of the plant is affected, a bacteriostatic solution based on copper or liquid Cryptonol can be applied to the affected area. The soil close to the collar should not be too wet.

Rotten fruit should be destroyed.

Superficial root proliferation caused by **Agrobacterium rhizogenes** is occasionally found on cucumber roots. The affected plants lack vigour and their yield is reduced. Steam sterilisation of the soil will reduce the incidence of these bacteria about which little is known at present.

FUNGI

Factfile 5

Cladosporium cucumerinum Ellis & Arth.

Causes: **Gummosis or grey anthracnose**

SYMPTOMS

Refer to **108, 112, 113, 117, 118, 125, 126, 127, 128, 133, 134, 258, 305, 318, 319, 320, 332, 333, 379, 380, 423.**

MAIN CHARACTERISTICS

● **Frequency and economic importance:** This fungus is widely spread over the colder, wetter regions of North America and Europe. It is also found in certain parts of Africa and south and east Asia. It affects numerous cucurbits. In France and the UK all cultivated varieties of cucumber are resistant. However, there are sometimes serious outbreaks on courgettes and melons. Outbreaks may occur in plant-raising nurseries as well as in the field.
● **Source:** The fungus lives on fruit and plant debris remaining on and in the soil, and also on seeds. It can survive on the structure of glasshouses where young plants are raised.
● **Spread:** The spores are dispersed and carried by wind, air currents, tools, workers' clothing and insects. Spores are fairly resistant to environmental conditions and will survive transport over long distances.
● **Conditions encouraging development:** *Cladosporium cucumerinum* tolerates cold, wet conditions often associated with badly drained land. Its optimum temperature range for development lies between 5°–30°C. The optimum for spore germination and penetration of the mycelium is about 17°C. Night temperatures of 15°C and day temperatures of 25°C are also very favourable. Penetration may take place after an overnight 6-hour period, or three close but distinct 2-hour periods, during which humidity is at saturation point. The disease develops rapidly, given 30 hours of 100% humidity. Its vigour decreases as the temperature rises above 22°C and it is hardly active at 30°C. After heavy rainfall, symptoms will appear on leaves and fruit in 3–5 days and sporulation takes place one day later.

CONTROL

● **During cultivation:** As soon as the first symptoms appear, the structures where the young plants or crops are growing should be aerated or ventilated to reduce humidity. It is also advisable to raise the temperature, especially during the night. Irrespective of the crop grown, the greatest care should be taken to avoid the presence of water on the surface of plants. If sprays have to be used, these should be applied during the morning.

Fungicides° should be applied as soon as possible, using the following active substances: maneb, mancozeb (zinc-metiram), chlorothalonil, dichlofluanid, benomyl, thiophanate-methyl, fenarimol and triforine. The last two active substances also have valuable anti-mildew properties.

Badly affected plants and fruit should be destroyed. After the crop is harvested, the plant debris and especially the fruit should be cleared.
● **Following crop:** If the outbreaks have affected young plants grown under cover, the equipment and the structure must be disinfected. Particular attention should be paid to the management of the nursery (see Appendix I, page 257): the plants must be kept warm; it is essential that humidity is kept at a low level.

IMPORTANT NOTE

° Approval of fungicides may vary from country to country and local labels should be consulted before use.

Preventive treatments should be applied using one of the fungicides mentioned above.

Where crops are grown in the traditional manner seeds harvested this year can be disinfected for 2 minutes in a 2% solution of sodium hypochlorite.

It is a good idea to rotate crops with non-susceptible plants, such as maize.

When the weather is cold and wet, it is advisable to keep a very careful watch on plants and to carry out preventative fungicidal treatments, particularly to young plants.

Among the Cucurbitaceae, only commercial varieties of cucumber are resistant.

Colletotrichum lagenarium (Pass.) Ellis & Halsted = *Glomeralla cingulata* (Stonem) Spaulet & Chenk var. *orbiculare* S.F. Jenkis & Wistead

Causes: **Anthracnose**

SYMPTOMS

Refer to photos 114, 115, 119, 120, 121, 122, 123, 124, 135, 136, 137, 255, 259, 260, 305, 317, 322, 334, 335, 336, 337.

MAIN CHARACTERISTICS

● **Frequency and economic importance:** Spread of this fungus is worldwide, but it is mainly active in countries with a wet climate and/or in those where little use is made of fungicides. In France and the UK it has become a 'curiosity' that can sometimes be seen in private gardens, mainly on melons and water melons.
● **Source:** Plant debris, particularly straw-based manure. On these materials, the fungus may persist in the soil for up to 5 years. It will also survive on seeds. As soon as contamination takes place, the fungus penetrates through the leaf cuticle.
● **Spread:** The conidia are formed in abundance on the acervuli, and are easily disseminated by water splashes and drips after rainfall, overhead spraying or condensation. It can also be spread by workers during the course of cultivation, and by certain insects. The spores germinate at temperatures between 5°–30°C.
● **Conditions encouraging development:** Infection often occurs after a wet period: 24 hours with the humidity at 100% and temperatures between 19°–24°C provide the most favourable conditions. Symptoms will appear in less than a week.

CONTROL

● **During cultivation:** If the outbreak occurs in a protected crop, the premises must be aerated and ventilated. In the field, avoid overhead watering. If this is not possible, water in the morning, so that the foliage can dry out quickly during the day.

Badly affected plants, and especially fruit, should be removed and destroyed.

Treatments based on zineb, maneb, mancozeb, and chlorothalonil should be carried out, applying the fungicides every 10 days and again after rain. Following the final harvest, all crop debris should be cleared and destroyed.
● **Following crop:** In countries where it still causes crop yield losses, the following measures are advisable:
● Do not used contaminated seeds taken from diseased fruit;
● Try to use a crop rotation;
● Destroy any cucurbits which may be carriers of the fungus;
● Take preventive action in good time, using the products mentioned above.

Some varieties of cucumber and water melon are tolerant or resistant.

(Fungi attacking the leaves)

(Fungi attacking the leaves)

Botrytis cinerea Pers. = *Botryotinia fuckeliana* (De Bary) Whetzel

Causes: **Grey mould**

SYMPTOMS

Refer to 109, 165, 166, 167, 168, 169, 170, 254, 271, 277, 308, 339, 354, 362, 367, 368.

MAIN CHARACTERISTICS

• **Frequency and economic importance:** This fungus has a wide host range and is found in all parts of the world. In Europe at the present time it causes more damage to protected vegetable crops than any other fungus. It can attack and colonise a wide variety of hosts and may cause considerable losses (particularly via lesions and senescent tissue on which it establishes and which is ideal for its growth). Crops affected include lettuce, peppers, aubergines and tomatoes which are rotated with cucurbits or are grown near by. In the UK and France, it causes considerable damage to cucumbers grown under glass.
• **Source:** The organism survives in various forms: conidia, mycelia, sclerotia on plant debris, in or on the soil.
• **Spread:** In the open, rain and wind are means of spread; under protection, air currents, and in particular condensation drips, overhead watering or hand sprinkling all disperse the fungal spores over the crop and the entire structure.
• **Conditions encouraging development:** The development of *Botrytis cinerea* depends mainly on the ambient relative humidity, its optimum being 95%, but also the presence of free water on the plants is favourable for infection. The water comes from the drips which fall onto the plants; they are formed by condensation on the structure, on the heat screen or directly on the leaves, especially towards dawn. The duration of moist atmospheric conditions is a critical factor for the development of the fungus; it is not a question of temperature. Although the optimum temperature lies between 17°C and 23°C, *B. cinerea* can infect plants at much higher or lower temperatures.

There are other factors which influence the behaviour of *B. cinerea*:
• Quality of light: maximum sporulation takes place at wavelengths less than 345 nm;
• Excess of nitrogen in the fertiliser (nitrate nitrogen): by changing the size and wall structure of the plant cells, they are made susceptible to the fungus;
• Various forms of stress, e.g. resulting from water, heat or light, which induces over-vigorous or spindly plants to become more susceptible;
• The receptivity of the host: there is not a great deal known about this factor. Apparently the susceptibility of the host varies throughout its life, and it is much more vulnerable at some stages in its development than at others. This is particularly noticeable round about the time that the first fruit is harvested.

CONTROL

• **During cultivation:** At present this is a very difficult disease to control because numerous crops, especially those grown under protection, are affected by strains of *B. cinerea* which are resistant to the fungicides normally used against them. In addition, climatic conditions of protected crops provide an ideal environment for this pathogen. It is advisable to take the following measures to reduce the incidence of disease:
• Ensure maximum aeration and ventilation in order to reduce humidity; thermal screens tend to increase the latter and reduce light. As far as possible, ensure that standing water does not collect on the plants, because this is the one practical way of reducing infection. Although this is often a tedious job, the greenhouse staff must be made aware that this is one of the most effective measures. The foliage can be pruned to give better aeration if this becomes necessary;
• Remove axillary shoots at an early stage so that the pruning cuts are as small as possible and therefore less prone to disease. If there are cankered areas on the stem these can be painted with a thick fungicidal mixture (thiram and/or iprodione and/or benomyl). The addition of paraffin oil makes the mixture more effective;
• Clear the crop area of plant debris as soon as possible, in particular the diseased fruit and plants;
• Remove plants which are too vigorous or spindly;
• Treat the plants with fungicides from two different groups, using them alternately:

• **Group 1:** chlorothalonil, dichlofluanid, thiram.
Until now, *Botrytis cinerea* has not become resistant to these multi-purpose products, whose effectiveness is unfortunately somewhat limited. Very recently, strains resistant to dichlofluanid have been discovered in Crete. Some of these strains have also proved less susceptible to captan and chlorothalonil. So far the strains do not seem to have spread; they are best suppressed or prevented by the alternation of fungi-

cide treatments using the compounds already mentioned.

Group 2: The benzimidazoles (benomyl, carbendazim), the thiophanates, the dicarboximides or cyclic imides (iprodione, procimidone, vinclozolin). The first two, initially very effective, have included resistant strains (very active and persistent) which are still found in field work; it is advisable not to use these products. The dicarboximides, very effective when they first came into use, have also resulted in resistant strains (low activity and non-persistent), which are now frequently found in many glasshouses. It is advisable to stop using the dicarboximides temporarily, or to use them much less in glasshouses and alternate, or replace them, with products from Group 1.

A mixture of fungicides[*] (vinclozolin + thiram) that has been recently marketed, but is not yet approved for use, may give better protection. Another fungicidal mixture (diethofencarb + carbendazim), again not yet approved for use, is specific against benzimidazole- and thiophanate-resistant strains (diethofencarb showing a negative crossed resistance with these families).

These two specialist fungicides are at present being used with some success on vines; their use on cucurbits will depend on whether approval is granted for these plants. It is advisable to be sparing with them as isolates resistant to diethofencarb have already been reported in France and Israel.

● **Following crop:** Outbreaks in young plants are rare. To prevent them, do not plant too deeply and do not earth up too much. If in spite of these precautions damage occurs, apply an anti-botrytis fungicide locally to the collar (directed spray).

As soon as the foliage becomes dense and the plants vigorous keep a strict watch for Botrytis, especially when the weather is overcast and particularly when the first fruit is nearly ready for harvest.

IMPORTANT NOTE

[*] Approval of fungicides may vary from country to country and local labels should be consulted before use.

(Fungi attacking the leaves)

Didymella bryoniae (Auersw.) Rehm. = *Phoma cucurbitacearum* (Fr.) Sacc.

Principal synonyms:
— *Mycosphaerella citrullina* (C.O.Sm.) Grossenb
— *Mycosphaerella melonis* (Pass.) Chiu & Walker
— *Ascochyta citrullina* (Cester) C.O.Smith

Causes: **Gummy cankers on stems, black rot in fruit**

SYMPTOMS

Refer to 116, 163, 164, 169, 171, 172, 246, 247, 269, 272, 278, 346, 347, 348, 363, 364, 366, 369.

MAIN CHARACTERISTICS

• **Frequency and economic importance:** This fungus is widespread throughout the world, especially in tropical and sub-tropical regions. It affects many of the Cucurbitaceae (water melon, melon, gourd, cucumber) grown under protection or in the field, causing considerable damage. Today in France, the only conditions favourable for its growth occur in protected crops; it thrives in warm, wet climates and it is only these crops which are really at risk. It may affect melons, courgettes and gourds, but rarely causes much damage.

• **Source:** The fungus lives on and/or in the soil, on dryish, non-decomposed plant debris, and will survive for more than a year in the form of dormant mycelia. It is highly resistant to drought, and it will live on the structures of the glasshouse. When fruit is rotted, many of the seeds will be contaminated both outside and inside. Thus both the survival of the organism and the spread of the disease are assured.

• **Spread:** The primary inoculum often consists of pycniospores and ascospores; the latter are expelled from the perithecia as soon as the air becomes humid. *Didymella bryoniae* is a potential danger to the cucumbers grown in glasshouse the whole year round as the infections ascospores may be produced at any time. They are often released 3 hours after the wetting of the affected parts; the amount of available light does not appear to have a significant effect. They are then dispersed by air movement.

The pycniospores are wet spores which leave the pycnidia in the form of a yellowish mass. They are dispersed by splashes and trickles of water and may be spread to other plants by machines during cultivation.

• **Conditions encouraging development:** Damage from the fungus is particularly severe in plants which have developed damaged areas or which have been weakened by stress or attacks from other micro-organisms, pathogens or pests. Temperature and humidity are factors which may limit the spread of the fungus. It can grow and fructify at temperatures between 5°–35°C, the optimum on cucumbers being somewhere around 23°C. On water melons, the optimum temperature is slightly higher, about 24.5°C, while on the melon it is lower at 19°C. This host becomes much less susceptible as temperatures rise.

Humidity is definitely the most important factor. Plants cultivated in a dry atmosphere are normally less susceptible than those grown at a higher humidity. Infection is rare at relative humidities of around 60%. It becomes very serious when humidity reaches 90%, and especially when there is standing water on the plants. A film of water which remains in place for an hour will help the pathogen to establish. The organism can penetrate the cuticle or epidermis directly, or via pruning lesions.

CONTROL

• During cultivation: Effective control of this disease depends on initiating control measures as soon as the first symptoms appear. This is particularly important when crops are grown under protection, and the grower should be aware of the influence that the micro-climate has on the progress of the disease (and also of the importance of his good management). By heating and ventilating his glasshouse he will reduce the humidity and reduce the incidence of standing water on the plants. For example, if the night temperature is kept at a higher level when crops are almost ready to harvest, the disease is much less active. Growers are often reluctant to incur this extra cost. Only they can decide whether or not the financial outlay is justified.

Concerning field-grown crops, water should not be permitted to lie on the ground or on the plants for too long. The answer is to use less water but to water more often. Spraying should be carried out in the morning or during the day but never at night.

All affected fruit, foliage thinnings and disbudding deris should be cleared at once from glasshouses or fields.

If fungicides are to be effective they must be used relatively intensively, particularly under cover. The plants should be sprayed weekly with compounds based on benomyl, tolyfluanid, mancozeb, iprodione (or other cyclic imides, procimidone, vinclozolin), imazalil, triforine and chlorothalonil. It is essential to alternate

210

these fungicides with others having a different mode of action, because several countries have reported fungicide resistance, particularly to benomyl.

Recent studies carried out on several farms all over France have detected isolates of *Didymella* which are resistant to benomyl. This fungicide should be abandoned or used very sparingly. Stem lesions can be painted with a thick fungicidal mixture consisting of several of the active ingredients mentioned above. After the crop is harvested, all plant debris must be cleared from fields and glasshouses. In the field this debris can be buried deeply.

• **Following crop:** Where crops are grown in the traditional manner in the field, only healthy seed should be used. It must not be collected from partially rotted fruit or it will already be contaminated and very early outbreaks of disease can be expected. A crop rotation over at least 2 years should be used.

With protected crops, especially if soil-grown cucumbers are the only crop, the soil should be disinfected with steam or methyl bromide.

Plant debris lying close to the glasshouse either by accident or design can also be a source of the pathogen, mainly by ascospores which form on it as soon as weather conditions permit. Ground in the immediate vicinity of a glasshouse must be as carefully treated as that on the inside.

Partitions in the glasshouse should be sterilised with an aqueous solution of formaldehyde, sodium hypochlorite or a quaternary ammonium salt.

If preventive measures are to be effective, it is essential that the micro-climate in the glasshouse is controlled (eliminating water film on plants) and fertiliser use monitored.

Even though varietal resistance may have been demonstrated, there is not one variety of the Cucurbitaceae at present grown commercially which is resistant to this fungus.

Some melon varieties of the 'American cantaloupe type have a certain degree of resistance.

(Fungi attacking the leaves)

Sphaerotheca fuliginea (Schlecht.) Poll.
Erysiphe cichoracearum D.C.

Causes: **Oidium or Powdery Mildew**

SYMPTOMS

Refer to 111, 186, 188, 193, 194, 195, 196, 197, 198, 199, 261, 338.

MAIN CHARACTERISTICS

- **Frequency and economic importance:** These two fungi are found worldwide. Their vigorous colonising potential reduces the leaf's functional surface; while plant death is rare because of intensive chemical treatments, yield is reduced and quality is affected. In France, until very recently, it was known exactly which species of *Oidium* spp. infected Cucurbitaceae. *Sphaerotheca fuliginea* and *Erysiphe cichoracearum* are the only species to have been observed in the field. Each usually grows separately, although they are sometimes found in association on melons as well as on cucumbers and courgettes. In fact the situation is more complex: while all strains infect cucumbers, some cannot colonise melons and/or courgettes. These strains are examples of physiological specialisation.
- **Source:** The form in which these fungi survive is not fully understood. Some workers implicate the perithecia (the sexual spore stage), but their ability to establish the disease from one year to another has not yet been demonstrated; in addition, they rarely appear on leaf spots. The pathogens may be able to survive as conidia (*Oidium*), either on cucurbits grown as a late crop followed by an early crop thus ensuring continuity, or on volunteer plants which could act as hosts to one or other of the fungi.
- **Spread:** The wind in the open and air movement under protection disperses the conidia. Certain insects also act as vectors.
- Conditions encouraging development: we have only recently been able to differentiate between the two species. Confusion must have arisen in the past and therefore some biological data are probably inaccurate. Unlike many of the fungal pathogens of the fungi Cucurbitaceae, the mildews do not need a film of water on the leaves before they infect. In fact, if the conidia are in contact with water, they are inhibited to a certain extent, which could explain the checks in epidemics during wet spells. Temperature is not a limiting factor in their development which takes place between 10°–35°C, the optimum being around 23°–26°C. Their growth cycle is relatively short: about 7 days elapses between infection by the conidia and the appearance of mildew spots.

The distribution of the two species of mildew during the year, according to region and type of crop, shows that they probably have slightly different climatic requirements. Generally outbreaks of *Erysiphe cichoracearum*, which is more tolerant of high rainfall, occur earlier and more often in protected crops than in the field. *Sphaerotheca fuliginea* on the other hand is dominant in dry conditions: under well-ventilated glass and in the open during summer.

CONTROL

- **During cultivation:** Treatment by fungicides is the most popular method. There are many specific anti-mildew materials available to growers, which we have divided into two groups, according to whether or not they are likely to provoke the appearance of resistant strains.

  Fungicides* presenting little or no risk of inducing resistant mildew strains: micronised sulphur, sulphur dust or flowers of sulphur, pyrazophos, chinomethionate, dinocap.

  Fungicides* which present a risk of inducing less susceptible or resistant mildew strains: bupirimate, fenarimol, triadimefon, triadimenol + chinomethionate, triforine, myclobutanil.

  For example, the benzimidazoles (benomyl, etc.) can no longer be used because resistant strains are now very widespread. Studies abroad have shown that *Erysiphe cichoracearum* is less susceptible to bupirimate; *Sphaerotheca fuliginea* is apparently less susceptible in the field to triadimefon. Fungicides in the 'risk' class which are sterol biosynthesis inhibitors, should be used in moderation. It is advisable to mix them or to alternate them with products not belonging to this group, i.e. fungicides classed as 'no risk'.

  When the season is finished, all plant debris as well as the plants must be cleared and destroyed.
- **Following crop:** Where protected crops are grown, all surfaces should be disinfected with an aqueous solution of formaldehyde, sodium hypochlorite or a strong solution of an anti-mildew fungicide.

  Plants should be selected with care. It is possible that some plants sold to growers are already infected. They should be checked on delivery and treated suitably if they appear to be substandard.

IMPORTANT NOTE

* Approval of fungicides may vary from country to country and local labels should be consulted before use.

There are resistant varieties of melon and cucumber (see Appendix III); these were selected empirically at a time when it was not really known which were the pathogens responsible. In practice, the expected resistance is not always operational because of virulent strains of the pathogen. A variety resistant to *Sphaerotheca fuliginea* is not necessarily resistant to *Erysiphe cichoracearum*. Strains of *Sphaerotheca fuliginea* able to overcome some genetic resistance to this mildew have already been reported. Only one cultivar of water melon is susceptible to this disease, all the others being resistant.

Many fungi which are antagonistic to these midldews have been reported: *Stephanoascus* sp., *Verticillium lecani*, *Acremonium* sp., *Tilletiopsis minor*, *Ampelomyces quisqualis*, etc. In France, none of them have proved to be of use in practice for the biological control of these mildews.

(Fungi attacking the leaves)

Leveillula taurica (Lèv.) Arn. = *Oidiopsis taurica* (Lèv.) Salm.

Causes: **Internal Powdery Mildew**

SYMPTOMS

Refer to **187, 189.**

MAIN CHARACTERISTICS

● **Frequency and economic importance:** This fungus is quite widespread, particularly in the countries of the Mediterranean basin. It affects a wide range of plants, both cultivated and growing wild. Of the cultivated plants, the cucurbits are not the most susceptible.

● **Source:** As we have said before, the fungus has numerous hosts among the cultivated plants (Solanaceae, artichokes, cardoons, leeks, etc.) as well as wild plants, and can thus overwinter.
● **Optimum conditions for growth:** Not much is known on this subject; a relative humidity of 50–70% and temperatures between 20°–25°C appear to encourage growth.

CONTROL

● **During cultivation, and for the following crop:** Cucumbers are affected by this fungus in France, and then only rarely. There is no need for protective chemical treatments; but fungicides should be applied as soon as the first symptoms are observed, as this mildew is difficult to control. Use the same type of anti-mildew fungicides as used for other mildews affecting cucurbits.

Factfile 11

Pseudoperonospora cubensis (Berk. and Curt.) Rostw.

Causes: **Downy mildew**

SYMPTOMS

Refer to 3, 74, 110, 173, 174, 175, 176, 177, 178, 179, 180, 181, 182, 190, 209.

MAIN CHARACTERISTICS

- **Frequency and economic importance:** This fungus is found worldwide and often causes a great deal of damage. Because of its environmental requirements it was first found in the tropical and sub-tropical regions; now it has moved north. It affects many of the Cucurbitaceae, but the physiological strains exist in each country. In France, between 1971 and the 1980s it attacked only cucumbers. Now it also affects melons, and it is often a factor causing reduced yield in both hosts. It has never been seen in *Cucurbita* species or water melons.
- **Source:** It is not yet fully understood how the organism survives from one season to another, especially in more northerly regions. It may produce oospores, but these have been found so seldom that they do not appear to play an important role. They may be able to survive on affected plant debris. In warm growing areas the pathogen flourishes to the detriment of various self-sown and cultivated cucurbits. In more temperate regions, it may be that out-of-season crops of Cucurbitaceae ensure the pathogens' survival. In some countries, it is active all the year round.
- **Spread:** When humidity reaches 100% and in temperatures of between 20°–25°C a large number of conidiophores carrying numerous conidia appear on the undersides of leaves (less often on the upper side). The spores are easily dispersed by wind and air currents, splashes and drips of water after heavy rain or overhead watering. A hot, damp wind will carry the spores for long distances; with successive plantings the mildew may become established over a wide area spreading from a primary focus of infection in a single plot or growing area.
- **Conditions encouraging development:** Like many mildews, growth is favoured by the high humidity such as after fog, dew, rain and overhead watering. The presence of standing water on the leaves is necessary before infection can take place; this occurs in 2 hours, if the temperature is between 20°–25°C. It may occur at temperatures between 8°–27°C, the optimum lying between 18°–23°C. The fungus can easily withstand high temperatures; it may survive a day at 37°C. Its cycle is relatively short: 3–4 days after infection the conidiophores appear.

CONTROL

- **During cultivation:** In view of the rapidity with which this disease spreads and the amount of damage it causes, it is essential to move fast once the symptoms appear. Avoid overhead watering field-grown crops, especially at night and in the morning when dew is falling, because this extends the period during which the leaves will remain wet.

  Atmospheric humidity should be reduced for protected crops to minimise the amount of free-standing water on the plants. Aeration and ventilation are essential in this type of cultivation. Sprinkling should not be allowed.

  Apply chemical treatments as soon as possible. There are many fungicides* to choose from, varying in their effectiveness according to the particular conditions. They can be separated into two groups:

- Contact fungicides, such as zineb, propineb, maneb, mancozeb (methiram-zinc), chlorothalonil, dichlofluanid, etc.
- Systemic or translaminar fungicides, with cymoxanil, phosethyl-A1, propamocarb HCl, oxadixyl, metalaxyl, etc.

At first it is better to use a systemic fungicide (often in combination with one or more contact fungicides), on a regular basis but at least once a week especially if conditions remain favourable (e.g. dew, rain, fog). Following this, contact fungicides can be used. Metalaxyl and acylalanines should be used with care as strains resistant to fungicides belonging to this chemical family have been found in several countries. It is essential to alternate them with active ingredients belonging to other chemical families.

Debris must be cleared and destroyed.

- **Following crop:** When glasshouses are badly contaminated, it would be advisable to disinfect all the surfaces and the soil.

  In the field, where the land is badly drained or retains a great deal of water, drainage should be improved.

  Throughout the growing season it is advisable to be aware of any warnings of outbreaks of this disease in your vicinity, and to monitor the

IMPORTANT NOTE

* Approval of fungicides may vary from country to country and local labels should be consulted before use.

(Fungi attacking the leaves)

crops very carefully to detect the first symptoms as early as possible.

Preventive chemical treatments should be applied especially in those areas where the disease recurs annually.

Apparently varieties of cucumber and melon which are oidium resistant are slightly less susceptible to mildew. Some cucumber varieties are highly resistant; in France there are a very few tolerant hybrids marketed. Resistant parent plants of the melon do exist; work is being carried out now on the introduction of resistance into Charentais melons. Resistant parent plants of the water melon have also been introduced.

Corynespora cassiicola (Berk. & Curt.) Wer
synonym: *Cercospora melonis* Cooke
Ulocladium atrum Preuss
 Ulocladium cucurbitae (Letendre & Roumeguere) Simmons
Alternaria cucumerina (Ellis & Everh.) Elliott (Alternaria)
Cercospora citrullina Cooke
synonym: *Cercospora cucurbitae* Ell. & Ev. (Leaf spot)
Septoria cucurbitacearum Sacc. (Leaf spot).

Causes: **Various leaf spots**

SYMPTOMS

Refer to 155, 156 (*Ulocladium atrum*), 157, 158 (*Corynespora cassiicola*).

MAIN CHARACTERISTICS

• **Frequency and economic importance:** All these fungi affect a wide variety of cucurbits. *Cercospora citrullina* is firmly established in tropical and sub-tropical regions. It does not appear to be present in Europe. *Ulocladium atrum* has recently become adapted to cucumbers; it is seen only occasionally in England and France and is no threat to crops. However, a *Ulocladium* morphologically similar but named *Ulocladium cucurbitae* has been known to cause severe damage in the USA. *Alternaria cucumerina*, *Septoria cucurbitacearum*, *Corynespora cassiicola* are all more widely spread throughout the world. *Corynespora cassiicola* is found sporadically on cucumbers, but usually it causes little damage.
• **Source – Spread – Conditions encouraging development:** We know very little about *Ulocladium atrum*. The fungus is definitely able to survive on crops as a saprophyte as can many *Alternaria* and *Stemphylium* spp. (very close relations). We have seen it quite often on cucumbers grown under protection where conditions are humid and relatively warm. In this environment *Ulocladium* grows rapidly and there is barely a week between infection and the appearance of the first symptoms. Numerous black conidia are quickly produced on the spots. They are dispersed by air movement, splashes and drips of water after sprinkling, spraying or heavy condensation.

Corynespora cassiicola infects many hosts (tomatoes, aubergines, etc.) and survives for at least 2 years on the plant debris and sometimes on the seeds. As with *Ulocladium atrum*, the incubation period is about a week. It will also develop in very humid conditions (especially under cover or in the field in hotter regions) and at temperatures between 25°–36°C. The conidia are dispersed and carried by the wind and air current under cover.

The spores of *Alternaria cucumerina* survive for several months in dry conditions but rapidly lose their viability in soil. Often the fungus survives on plant debris, as dormant mycelium. Infection occurs when the humidity rises. Standing water on the leaves (following dew falls, rain, spraying) and temperatures between 21°–32°C are determining factors in the development of the disease. Incubation lasts from 3–12 days depending on climatic conditions.

The spores formed on the damaged tissues are very easily dispersed and carried by wind and air currents.

Cercospora citrullina survives on plant debris from one season to the next, and it also appears to live on the seeds of certain cucurbits. Tropical climates (heat and high humidity) favour its development, which is extremely rapid when temperatures are between 26°–32°C and there is standing water on the leaf surfaces. Under these conditions, the fungus sporulates freely and the conidia can be carried over long distances by humid winds. Contaminated tools, rain or irrigation will also disperse the fungus.

Septoria cucurbitacearum, like the fungi described above, thrives on high humidities; conversely, its optimum temperature is quite low, between 16°–19°C. It is able to survive in the soil on plant debris. The conidia produced on the leaf spots are probably spread by water splashes, heavy dews, etc.

CONTROL

• **During cultivation:** As soon as the first symptoms appear, it is advisable to ensure that there is no standing water on the plants and that watering is not too frequent. Under cover, ventilate as much as possible and do not allow the use of sprinklers or sprays.

In the field, overhead irrigation should be avoided; if it is absolutely necessary, it should be carried out in the morning so that the foliage has a chance to dry off quickly; irrigation should never be carried out at night. Badly affected plants and leaves should be removed and destroyed.

Many fungicides* offer good protection, for example:
• chlorothalonil, maneb and mancozeb, effective enough against all five of these fungi, are particularly active against *Septoria cucurbitacearum*;
• iprodione and imazalil, useful especially for *Alternaria cucumerina* and *Ulocladium atrum*;
• the benzimidazoles, in particular benomyl, will control *Corynespora cassiicola* and *Cercospora citrullina*. Although there are not yet many confirmatory results, certain fungicides belonging to the sterol biosynthesis inhibitors

---

IMPORTANT NOTE

* Approval of fungicides may vary from country to country and local labels should be consulted before use.

group are reported to be effective against these two fungi. To control *Corynespora cassiicola* benomyl should be used alternately with other fungicides as resistant strains are beginning to be reported.

When the season is over, it is advisable to clear and destroy the crop residues.
• **Following crop:** After a serious outbreak of *Corynespora cassiicola* in protected crops, the area should be disinfected. If the fungus appears each year, preventive chemical treatments should be applied early in the season. There are numerous cucumber varieties which are resistant to this fungus. The lines resistant to *Corynespora* are also resistant to *Ulocladium cucurbitae*.

With regard to other fungi, disinfection of the seeds or treating them with one or several fungicides has proved to be effective, as well as crop rotation for field-grown crops. Early diagnosis and prompt chemical treatment should ensure that fungal growth is restricted.

Fungi causing fruit rot

A great many fungi cause fruit rot of cucurbits. However, this section is concerned only with the fungi which have not been described in other Factfiles. In particular, we shall deal with **Choanephora cucurbitarum** and **Rhizopus nigricans**.

The main species of fungi named in the literature as agents of fruit rot are as follows:

Alternaria tenuis Auct.
Botrytis cinerea Pers. = Botryotinia fuckeliana (de By.) Whetz.
Choanephora cucurbitarum (Berk. & Rav.) Thaxt.
Cladosporium spp.
Diplodia natalensis Pole-Evans = Diplodia gossypina Cke. (A.) = Physalospora rhodina (Berk & Curt.) Cke (S)
Fusarium spp.
Geotrichum candidum Link = Oospora lactis (Fr.) Lind.
Macrophomina phaseoli (Tassi) G.Goid.
Penicillium spp.
Phytophthora spp.
Pythium spp.
Rhizoctonia solani Kühn = Thanatephorus cucumeris (Frank) Donk.
Rhizopus nigricans (Fr.Ehr.) = Rhizopus stolonifer (Fr.Ehr.) Wuill
Sclerotinia sclerotiorum (Lib.) de By.
Sclerotium rolfsii Sacc. = Athelia rolfsii (Curzi) Tu & Kimbrough.
Stemphylium botryosum Wallr = Pleospora herbarum (Fries) Rabenhorst
Trichothecium roseum (Lk.) Fr. = Cephalosporium roseum Cda.
Ulocladium botrytis

SYMPTOMS

Refer to 345, 356, 357, 373, 374, 375, 376.

MAIN CHARACTERISTICS

- **Frequency and economic importance:** many of these fungi affect mainly fruit of field-grown cucurbit crops. The number of outbreaks varies from one growing area to another and from one country to another. Some are established in hot tropical regions, others in more temperate zones.

It should be noted that many varieties can live only in association with other fungi or bacteria (Erwinia spp., Bacillus sp., Pseudomonas sp.).

Choanephora cucurbitarum is usually found in tropical countries, especially on gourds. It infects the wilted corollas before colonising and rotting the fruit. Rhizopus nigricans is found further north; it affects very many types of ripe fruit. Choanephora cucurbitarum is not found in France but Rhizopus nigricans causes considerable damage there, reducing the yield of marketable fruit.

- **Source:** Most of these fungi are saprophytes, able to live in soil on any type of plant debris. Each of them has its own specific structure which enables it to survive in soil: chlamydospores, microsclerotia, sclerotia, etc.

Choanephora cucurbitarum and Rhizopus nigricans are highly developed saprophytes. Their enzymic activity enables them to degrade a large number of substrates in vegetable matter and thus to survive in the soil from one year to another. They also have a wide host range especially Rhizopus nigricans. As well as cucurbits it will invade many other hosts: peas, peppers, avocado pears, tomatoes, figs, strawberries, etc. By means of their chlamydospores and zygospores they can survive for many years in the absence of a host.

- **Penetration:** This is usually via various lesions (growth splits, sun burn, apical necrosis, insect punctures, mechanical damage, etc.) on the parts of the plant which may or may not be in contact with the soil. Some fungi will penetrate directly through the cuticle.

- **Spread:** Often the fungi spore copiously on the fruit, and the spores are then dispersed by wind, rain and watering, resulting in splashes spread by spore directly or on particles of soil. Some contaminate the seed and/or are carried by insect vectors (flies, bees, etc.). When these insects visit the rotted fruit for its sugar they pick up numerous spores which are later dispersed.

- **Conditions encouraging development:** The type and stage at which decay becomes apparent depends largely on the conditions. Decay is particularly rapid after heavy rainfall. Water remaining on the fruit, or between the fruit and the ground, will encourage rapid deterioration. Moreover, heavy rain coming after a period of drought may cause numerous growth splits which provide entry points for the fungi. Badly managed irrigation can have the same effect.

These micro-organisms find it less easy to invade unripe fruit, but cause a great deal of damage to over-ripe fruit.

Choanephora cucurbitarum and Rhizopus nigricans are not aggressive pathogens. Before they can invade their hosts they need to colonise a source of food such as wilted flowers or wounds (damage from insects or machines etc.). Senescent tissue is well colonised. Their optimum temperatures differ. Choanephora cucur-

bitarum, established in tropical regions, thrives in a hot, humid climate and grows particularly well at temperatures of 25°C or over. *Rhizopus nigricans* also likes high humidity but its optimum temperature is very much lower. It is very aggressive at 15°C, but its pathogenicity declines above 20°C. Above 37°C it does not appear to be viable.

CONTROL

● **During cultivation and the following crop:** These are difficult fungi to eradicate; it is advisable to adopt the following measures:
● Irrigation should be regular: do not leave the plants without water for a lengthy period and then over-water to compensate. When spraying, do not use too much water and never spray at the end of the day. If these fungi appear on crops grown under protection, it is essential to restrict the ambient humidity and to provide maximum aeration.
● Fruit should be unblemished and not left to become over-ripe. Rotted fruit must be removed during and at the end of the season.
● It is rarely necessary to apply specific fungicides. Often they are not really effective, because the fruit is more or less covered by flowers and therefore not fully exposed to the spray.
● When the next crop is growing, ensure that the fruit is not in contact with the ground, and that there are no puddles of water close to the fruit.

Various Pythiaceae (*Pythium* spp., *Phytophthora* spp.)

Cause: **Damping-off, Root rot and root loss, Collar canker**

The Cucurbitaceae are vulnerable to attack from several species of *Pythium* and *Phytophthora*; those most frequently reported are as follows:

Pythium aphanidermatum (Edson) Fitzpatrick.
Pythium debaryanum Hesse (Syn. *Pythium irregulare*)
Pythium ultimum Trow.
Pythium butleri Subramanian
Pythium intermedium de Bary
Pythium irregulare Buisman
Pythium splendens Braun
Phytophthora capsici Leonian
Phytophthora cryptogea Petherbridge & Lafferty
Phytophthora drechsleri Tucker
Phytophthora melonis

SYMPTOMS

Refer to 129, 130, 131, 132, 208, 223, 226, 227, 230, 231, 232, 233, 262, 343, 359, 370, 371, 372, 419, 420, 421.

MAIN CHARACTERISTICS

● **Frequency and economic importance:** It is known that the Pythiaceae are scattered throughout the world, some being active only in tropical zones while others are more widespread. We will deal with them all in the same Factfile as their symptoms are similar; they often develop best under the same conditions and control methods are identical. All cucurbits are susceptible to one or more Pythiacious fungi, which can cause considerable damage both to seedlings and adult plants.

● **Source:** The organisms can survive as soil saprophytes on organic matter. They can attack many hosts besides cucurbits, thus ensuring their multiplication and survival. They live in soil in the resistant form of oospores, chlamydospores, and to a lesser degree, sporangia. Some oospores have been known to survive from 2 to 12 years.

● **Spread:** These are aquatic fungi, perfectly adapted to living in water films in soils and soil-less cultures. They are not only spread by water: substrates, seeds and plants may be contaminated and introduce the fungi into the growing system. Some can also be aerially dispersed, especially *Phytophthora capsici*, by splashes resulting from overhead watering or heavy rain.

● **Conditions encouraging development:** The Pythiaceae are not all equally pathogenic, and although they can attack plants, various factors favour them:

● The presence of water is nearly always essential. In a very wet soil the reduced gaseous exchanges are an ecological advantage for these fungi, to the detriment of other fungi and microorganisms which may be competing for the organic material available in the soil;

● Temperature influences the behaviour of these fungi in different ways. Some, such as *Pythium ultimum*, do well in cold soils at temperatures around 15°C; others have higher optimum temperatures. This is certainly the case with *Pythium aphanidermatum*, now found in the fairly warm substrates of soil-less cultures, whereas before it was normally confined to more tropical latitudes.

The main species found in France are, as far as we know, *Pythium ultimum*, *Pythium aphanidermatum*, *Pythium irregulare* and *Phytophthora capsici*; their temperature range is shown in the table below:

	P. ultimum	*P. aphanidermatum*	*P. capsici*
Optimum temperature	15°–20°C	26°–30°C	24°–26°C
Temperature range	2°–42°C	5°–41°C	8°–37°C

(Fungi [& nemotodes] attacking mainly roots, collar and stem)

- The host does not remain equally susceptible throughout its life. Young plants are often highly susceptible and adult plants may become so, if they suffer stress from environmental conditions or agricultural operations. For example, plants which have sustained root loss, or which are carrying too many fruit (especially young plants), may become extremely susceptible to attacks from *Pythium*, especially in systems of soil-less culture.

CONTROL

The methods described here can be used to control outbreaks of any or all of the Pythiaceae on cucurbits.

- **During cultivation:** It is advisable to check, and if necessary adjust, temperature and irrigation. Is the temperature high enough? Are you watering too much?

The roots and collar of the plants should be treated with appropriate fungicides as soon as possible. There are various possibilities, depending on the crop and the growing method:
- All of the substrate in the nursery frames should be drenched with a fungicidal solution as a matter of urgency.
- For a soil-based crop, without localised irrigation, the fungicide should be sprayed on the bases of the diseased plants and the adjacent healthy plants.
- For soil-based or soil-less crops, where the water and/or nutrient solutions are applied locally direct to the plant base (drip system), the fungicide can be applied via the irrigation water. However, in certain substrates, this method of application may mean that it is not evenly distributed throughout the root systems.

Several fungicides° are in common use:
- metalaxyl

IMPORTANT NOTE

° Approval of fungicides may vary from country to country and local labels should be consulted before use.

- furalaxyl
- propamocarb HCl
- etridiazole

It is difficult to recommend the best dosage for each of these products, because this differs according to the method of cultivation. In soil-less cultures, especially in NFT, and in substrates such as rock wool, the amount of active ingredient should be reduced to avoid any danger of phytotoxicity. It is also advisable to alternate with products belonging to different chemical families with a different mode of action, as we have recently found isolates of several *Pythium* species resistant to metalaxyl and furalaxyl.

Dead plants must be destroyed as soon as possible.

Outbreaks on leaves and fruit can be treated with sprays. This may not be effective on unstaked plants, because the spray often does not reach the underside of the fruit. Do not allow the fruit to lie in water or the soil to remain wet for too long.
- **Following crop:** Use treated compost in the seedling frames; sand and/or soil dug from the field should not be mixed with the rotting compost, because it could be contaminated.

The pots should not be placed directly on the soil, particularly if the latter has not been disinfected; it is better to put them in rows on shelving or on plastic sheeting. If the plastic is covered by a layer of peat, this should be renewed frequently.

Planting out: Avoid setting the plants into soil or a substrate that is too cold. Also avoid overwatering round the collars of the plants; this will ensure that the root system has the best chance of becoming established.

In soil-less culture, contaminated substrate should be replaced or disinfected.

Check the source of your irrigation water: if it comes from canals or pools, it may well be contaminated.

Rhizoctonia solani Kühn = *Thanatephorus cucumeris* (Frank) Donk.

Causes: **Decay in Roots and Collar**
Canker on stems and fruit

SYMPTOMS

Refer to **224, 234, 235, 236, 237, 263, 344.**

MAIN CHARACTERISTICS

• **Frequency and economic importance:** This fungus is present in almost all soils in which vegetable crops have been grown and it can invade many hosts. Cucurbits, particularly the melon, like other market garden plants, are susceptible. It attacks both young plants (damping-off) and mature plants, especially those which are soil grown. When the roots are affected, the symptoms frequently go unnoticed or are incorrectly identified. This fungus tends to be underestimated, although it can cause very considerable damage.
• **Source:** The fungal mycelia or sclerotia survive perfectly well in soil. This pathogen has a wide host range (known hosts exceed 25) and can invade and live on all types of plant debris (salad vegetables, tomatoes, aubergines, peppers, beans, several species of weed, etc.).
• **Contamination:** This can occur via horticultural substrates or potato crops previously grown on the land.

• **Conditions encouraging development:** It is extremely common in the soil of market gardens in which leguminous crops have previously been cultivated for several consecutive seasons. It thrives at temperatures between 15°–26°C in wet, heavy soils and in drier and lighter soils.

CONTROL

• **During cultivation:** When an outbreak of *Rhizoctonia* affects a growing crop, treatment is rarely necessary, unless there are serious outbreaks during propagation or soon after planting out. Quintozene (PCNB) or thiophanate-methyl can be applied locally to the plant base, possibly without much effect.
    Diseased plants and plant debris must be destroyed both during and after the growing season. Too much watering should be avoided, especially on heavy soils.
    Fruit decay is usually not sufficiently serious to justify treatment but fungicidal application to fruits is often ineffective.
• **Following crop:** It is essential to use pathogen-free substrates during propagation and not to place the plants directly on the soil. Where the soil is grossly contaminated, it must be treated either with a fumigant such as dazomet or methan sodium, or by applying quintozene to the soil, or by heat treatment to the soil. Quintozene is becoming less and less effective and will very soon be replaced with *mepronil* and *pencycuron* which should give better results.
    In hot countries, solarisation (see Factfile 17) is an economical and effective method which will eliminate this fungus, living as it does in the surface layers of the soil.

(Fungi [and nemotodes] attacking mainly roots, collar and stem)

(Fungi [and Nematodes[attacking roots, collar and stem)

Phomopsis sclerotioides Van Kesteren

Causes: **Black Root Rot**

SYMPTOMS

Refer to 201, 204, 220, 221, 228, 229, 240, 241, 242, 243, 244.

MAIN CHARACTERISTICS

• **Frequency and economic importance:** This fungus is found in Canada, Malaysia and several North European countries such as Denmark, Germany, England, the Netherlands and France. In these countries it mainly affects cucumbers and melons. Its incidence in intensive systems of production has forced growers to treat the soil or to switch to soil-less systems. Other cucurbits are not seriously affected, although a great many species in the family are susceptible to a greater or lesser extent, with the genus *Cucumis* being the most susceptible.

• **Source:** The numerous pseudo-sclerotia and pseudo-stromata formed by the fungus on the decaying roots allow it to survive in the soil for several years. It appears to become established in the roots via the root hairs.

• **Transmission and spread:** This soil fungus can rapidly colonise soil; if the latter has been treated, the process is even speedier. It is also transmitted via plants, especially if these have been in contact with contaminated soil. It is possible that soil dust may disperse the pathogen, partially in soil-less systems, while tools and machines can carry the organism via any lumps of earth adhering to them.

• **Conditions encouraging development:** These are not well defined. The fungus appears capable of adapting to fairly cool soils: at 10°C it causes numerous brown lesions. At 20°C damage is relatively slight and it tends to form pseudo-stromata and pseudo-sclerotia, which ensure its survival.

CONTROL

• **During cultivation:** It is almost impossible to apply effective treatment at this stage. It is, however, advisable to get measures underway which will both restrict the development of the pathogen and improve the plants' growing conditions, such as:

• Earth-up the plants so as to encourage new root production to supplement the old necrotic roots. If parts of the plant above ground have been affected (on peat or pozzolano + peat, which is less common), sterile peat can be spread round the collar to allow additional roots to grow;

• Damp the plants during the hottest part of the day to avoid too much transpiration leading to the desiccation and death of the plants;

• Monitor irrigation closely; then if the plants are wilting you will know that the cause is *Phomopsis sclerotioides* and not lack of water. Growers very often mistakenly increase irrigation; this can lead to root death.

Applications of fungicidal* solutions (based on benomyl, carbendazim and thiophanate-methyl) to the collar of the plants or by means of a drip irrigation system, gives satisfactory results. There is a risk of phytotoxicity if the amount of active ingredient is too high or the treatments are too frequent.

During and at the end of the crop pull up the affected plants carefully and remove and destroy as many as possible of the diseased roots.

• **Following crop:** A successful crop rotation is difficult to achieve as this fungus survives in the soil for a very long time, even though non-susceptible crops are grown. When melons and cucumbers are grown under protection, soil disinfection could be considered. Steam can be used, although not many farms are equipped for this yet; methyl bromide* and/or chloropicrin (in countries where these are authorised) are the most effective products. Dazomet and metham sodium are less effective. Soil treatment is effective providing that it is carried out correctly and for every crop.

IMPORTANT NOTE

* Approval of fungicides may vary from country to country and local labels should be consulted before use.

Grafting onto resistant or tolerant rootstocks is an alternative. The rootstocks selected will depend on the crop:

• For melons, F1 hybrids should be used, produced by crossing *Cucurbita maxima* and *Cucurbita moschata*;

• For cucumbers, *Cucurbita ficifolia* is fairly tolerant to *Phomopsis sclerotioides*; it is also resistant to *Verticillium dahliae* as well as *Pythium* spp., which cause root rot.

(Fungi [and Nematodes] attacking roots, collar and stem)

Pyrenochaeta lycopersici Gerlach.

Causes: **Corky Root**

SYMPTOMS

Refer to **203, 238.**

MAIN CHARACTERISTICS

● **Frequency and economic importance:** This fungus is not widespread. It has been reported on tomatoes in the USA, the Mediterranean basin and several European countries. It can also attack green salad crops, cucumbers and melons. In France the fungus affects tomatoes and (very rarely) lettuce, and there are also sporadic outbreaks on melon plants, although these are not economically important.

● **Source:** The pathogen can penetrate deep into the soil; it can colonise the roots of cultivated plants such as tomatoes, lettuce, aubergines, haricot beans, and several weed species. It can survive for several years in the soil, mainly on roots, in the form of chlamydospores. Monocropping tomatoes will increase the inoculum levels in the soil.

Strains described as cold, temperate and warm can exist in the same soil, each becoming active at the appropriate temperature.

Growth of the fungus in soil is very slow.

● **Spread:** Via all substrates and agricultural implements.

● **Conditions encouraging development:** In the soil, optimum development temperatures are from 15°–20°C for the northern European strains. There are various strains whose optimum temperatures vary according to their source. Strains originating in the Mediterranean basin (Tunisia, Libya) are pathogenic at temperatures of 26°–30°C.

CONTROL

● **During cultivation:** There is no way to eliminate the pathogen once it has colonised the roots, without damaging the roots. The answer is to try and keep the plants alive as long as possible by the following means:

● Earth-up the plants to encourage new root production to supplement the old diseased roots. If crops in soil-less culture are affected (on peat or pozzolano + peat), peat can be spread around the collar to encourage additional root development.

● Damp the plants during the hottest part of the day to avoid too much transpiration leading to the desiccation and death of the plants;

● Monitor irrigation closely; if the plants are wilting you will know that the cause is *Pyrenochaeta lycopersici* and not lack of water. Growers very often mistakenly increase irrigation; this can lead to root death.

Applications of fungicides[*] (based on benomyl, carbendazim and thiophanate-methyl) to the collar of the plant or via a drip irrigation system are sometimes recommended for field-grown crops. The effectiveness of these measures is questionable.

When the cropping season is over, it is advisable to remove the plants with care so that as many diseased roots as possible may be destroyed.

● **Following crop:** A crop rotation is difficult to achieve as this fungus survives in the soil for a very long time, even though non-susceptible crops are grown. When melons are grown under plastic, soil disinfection may be worthwhile.

IMPORTANT NOTE

[*] Approval of fungicides may vary from country to country and local labels should be consulted before use.

Steam can be used, although not many farms are equipped for this yet; methyl bromide[*] and/or chloropicrin (in countries where these are authorised) are the most effective products. Dazomet and metham sodium are also very effective.

Solarisation is a possibility; results have been quite striking in some Mediterranean countries. The technique consists of covering the affected soil (well prepared and wetted) with a polyethylene film, 25–40 microns thick, and keeping it in place for at least a month during the sunniest period of the year.

(Fungi [and Nematodes] attacking mainly roots, collar and stem)

Fusarium solani f. *cucurbitae* Snyder & Hansen

Causes: **Collar rot (strain 1)**
       Fruit rot (strain 2)

SYMPTOMS

Refer to **127**.

MAIN CHARACTERISTICS

● **Frequency and economic importance:** This non-vascular Fusarium is fairly rare in France, although it is sometimes found on the collar of courgette plants, which are very susceptible. From time to time it causes serious outbreaks in the USA and Australia. It can infect other cucurbits such as pumpkins, water melons, melons, cucumbers, etc.
● **Source:** The organism can survive on plant debris in the soil for at least 2 years, in the form of chlamydospores (resting spores). Seeds may be contaminated both on the outside and under the seed coat, and the fungus can survive on seeds for at least 2 years. For this reason, they are often the source of the pathogen on farms.
● **Spread:** The organism is spread by seeds, and also by soil dust during cultivation (ploughing, soil levelling, etc.).
● **Conditions encouraging development:** Very little is known on this subject.

CONTROL

● **During cultivation:** There are no effective means of controlling this disease. Local applications of a benomyl or carbendazim-based fungicide to the collar of the plant could have some effect. Remove any diseased plants during and at the end of cultivation.
● **Following crop:** Where crops are field-grown, a rotation should be worked out: this *Fusarium* remains in the soil for no more than 3 years. Cucurbit species should not be included in the rotation.

With protected crops, soil disinfection can be considered, using for example methyl bromide. Whether or not this treatment is effective is questionable.

It is essential to use healthy seeds. They can be disinfected by treating them in water at 52°C for 15 minutes, but this method is not 100% reliable and it reduces the germination rate of the seeds.

Penicillium oxalicum Currie & Thom (formerly known as *Penicillium crustosum*)

Causes: **Blue Canker on stems**
       Stylar rot in fruit

SYMPTOMS

Refer to **248, 273**.

MAIN CHARACTERISTICS

● **Frequency and economic importance:** Outbreaks of this fungus on cucurbits are of recent occurrence; at the moment and as far as we know, it affects only maize, causing rotting of the ears and wilting and death of the young plants. Outbreaks in cucumber plants have been reported in Canada, England and the Netherlands. It was introduced into Europe in 1987–88. It can cause considerable damage; for example, in some glasshouses in Canada 50–60% of the plants may be affected. It is a new threat to protected cucumber crops.
● **Source:** At present very little is known about this fungus. Like many *Penicillium* species it appears to be capable of living in soil as a saprophyte, decaying a wide range of plant material.
● **Spread:** Under cover, the numerous spores present on the cankers are dispersed by air currents, splashes and drips resulting from sprinkling and overhead watering. It is easily spread during cultivation operations via knives and hands.
● **Conditions encouraging development:** The fungus appears mainly on plants suffering from stress: whether affected by climatic conditions or nutritional problems, these plants appear to

be more susceptible to infection by this pathogen. The protected environment is particularly favourable for its development: serious outbreaks occur after a series of hot, humid days and low night temperatures. It does not seem capable of infecting plants unless there is damaged tissue or a source of food. Numerous pruning and thinning cuts provide ideal sites for colonisation to begin.

CONTROL

● **During cultivation:** As soon as symptoms appear it is advisable to adjust the environment of protected crops; it is particularly important to minimise day/night temperature differentials. Humidity should be reduced by means of ventilation. Check that the nitrogen content of the fertiliser is not too high.

Spraying with fungicides based on benomyl, thiabendazole or iprodione is quite effective, but care should be taken not to over-use them. Resistance to these products has already been observed in some other *Penicillium* species. Alternate them with other fungicides such as chlorothalonil or imazalil which are just as effective.

The cankers should be painted with, for example, a thick solution based on thiram and benomyl.

During and at the end of cultivation dead plants and plant debris should be destroyed, taking care not to spread the pathogen. It is best to put the plants into plastic bags before removing them from the crop.
● **Following crop:** Before the next crop season, the frames and surfaces of the glasshouses, etc., should be disinfected with a formaldehyde solution, sodium hypochlorite or a quaternary ammonium salt. Satisfactory control of this disease is best achieved through good plant management, avoiding as far as possible climatic or nutritional stresses.

(Fungi [and Nematodes] attacking mainly roots, collar and stem)

(Fungi [and Nematodes] attacking mainly roots, collar and stem)

Sclerotinia sclerotiorum (Lib.) de By.

Causes: **Stem cankers and Fruit rot**

SYMPTOMS

Refer to **267, 268, 274, 275, 355, 365.**

MAIN CHARACTERISTICS

- **Frequency and economic importance:** This fungus is found worldwide and attacks a great number of hosts, particularly vegetable crops. Nearly all the cucurbits are susceptible.
- **Source:** The numerous large sclerotia or the mycelium present on and in plant debris left lying about on the ground will enable the fungus to survive for several years. It has a wide host range and can infect more than 300 different species, prominent among them being lettuce, tomatoes, aubergines and peppers, which are often in rotation with cucurbits.
- **Infection:** This occurs in one of two ways:
- Through the mycelium produced from the sclerotia; infection in this case is at soil level;
- Through ascospores produced by the apothecia (organs which intiate sexual reproduction in some fungi and which form from the sclerotia), which infect the above-ground parts of cucurbits and also ensure the dispersal of the pathogen over a distance of several hundred metres.
- **Conditions encouraging development:** The fungus grows best at relatively low temperatures, between 15°–18°C (minimum 5°C, maximum 30°C) and fairly high humidity, particularly at the collar level or under the foliage of the crop. It is very susceptible to carbon dioxide, which is why it is found only in the first few centimetres of soil. It develops best in light, humus-rich soils.

CONTROL

- **During cultivation:** The first priority is to remove dead plants with sclerotia and reduce the amount of water, as well as ensure maximum aeration and ventilation for crops grown under protection.

Overhead irrigation should be avoided as far as possible.

If the outbreak is serious, chemical protection should be considered; this will depend on where the symptoms are found:
- If the collar is affected, treat it with one of the following fungicides: benomyl, thiophanate-methyl, iprodione, vinclozolin, procimidone;
- If there are cankers on the stem, they can be painted with the same preparation described for controlling *Botrytis cinerea* (refer to Factfile 7).

If the above-ground disease incidence appears to be increasing, overall spraying with kidan (iprodione + neutral oil) can be carried out.

At the end of the crop season, it is advisable to remove and destroy infected plants together with their sclerotia. This should be done immediately at the end of the crop and is particularly important when green salad crops follow as they are also highly susceptible hosts to this fungus.
- **Following crop:** Many crops (lettuce, aubergines, haricot beans, tomatoes) which are grown in rotation with cucurbits are liable to be affected. Therefore, if the growing area is heavily contaminated the soil should be disinfected, or the crops grown somewhere else. Several fumigants* (methyl bromide, dazomet, metham sodium, etc.) are suitable, or steam can be used.

Where the soil is contaminated, particularly in glasshouses, etc., efforts should be made to eliminate any areas of standing water, because it is here that the apothecia are produced. These areas, especially where there are irrigation channels, can be covered over with plastic sheets; this will trap the ascospores beneath and reduce infection.

IMPORTANT NOTE

* Approval of fungicides may vary from country to country and local labels should be consulted before use.

Meloïdogyne spp. (gall nematodes)

Causes: **Root Gall**

SYMPTOMS

Refer to **222, 250, 251.**

Depending on the region and the country, different species of *Meloïdogyne* may be active:
- *Meloïdogyne arenaria*
- *Meloïdogyne javanica*
- *Meloïdogyne incognita*
- *Meloïdogyne hapla*

MAIN CHARACTERISTICS

- **Source:** These nematodes can persist in soil for more than 2 years in the form of egg masses protected by a viscous coating. They have a wide host range and attack many vegetable and fruit crops (artichokes, aubergines, peppers, cucumbers, melons, green salad plants, tomatoes, almond trees, peaches, olives, etc.).
- **Spread:** Via contaminated plants, tools and irrigation water.
- **Conditions encouraging development:** These organisms thrive best at fairly high temperatures (18°–27°C) often found in light, sandy soils, except for *Meloïdogyne hapla*, which can tolerate lower temperatures.

CONTROL

- **During cultivation:** There is no effective remedy; however, it is advisable to earth-up the plants and to water them during the hottest part of the day.

The roots of affected plants have vast numbers of nematodes and must be removed and destroyed.
- **Following crop:** Deep ploughing often repeated several times in the middle of summer will expose the nematodes to heat and they will dry out and die. While this reduces the nematode population in the soil, it is rarely sufficient on its own.

Soil disinfection is a common solution. Steam can be used, or more often methyl bromide* alone or in combination with chloropicrin or dichloropropane-dichloropropene (D.D.). There are other suitable fumigants such as dazomet or metham sodium.

It is possible to control this nematode by biological means. The method depends on certain fungi in the genus *Arthrobotrys* (in particular *Arthrobotrys irregularis*) being pathogens of nematodes; the method has been used for some years. The eradication programme has to be started at least 2 months before cultivation begins, especially in lightly infested soils. The fungus is scattered on the soil and lightly covered. The method is feasible only under certain conditions:
- Soil pH should exceed 6.4, the proportion of soluble salts should be less than 2%, and the proportion of organic matter greater than 8%;
- Plants must be healthy at planting;
- No fungicides should be applied to the soil.

Where the land is badly contaminated, the soil should be disinfected before introducing the fungus.

Melons can be grafted onto *Benincasa*; this rootstock is resistant to these nematodes.

IMPORTANT NOTE

* Approval of fungicides may vary from country to country and local labels should be consulted before use.

(Fungi [and Nematodes] attacking mainly roots, collar and stem)

(Vascular fungi)

Fusarium oxysporum f.sp. *cucumerinum Owen*

Causes: **Fusarium Wilt of Cucumbers**

SYMPTOMS

Refer to 257, 281, 284, 294, 295, 296, 297.

MAIN CHARACTERISTICS

● **Frequency and economic importance:** This fungus is not widespread. It has been found in various countries, notably England and the Netherlands. We have seen it in China, and it is found also in Japan. So far it has not appeared in France. When an outbreak does occur, unlike verticillium wilt it affects a large number of plants.

● **Source:** Like the other *Fusarium oxysporum* species, this *Fusarium* is able to survive in soil for many years in the form of thick walled chlamydospores, that are able to survive adverse environmental conditions.

● **Spread:** Cankers form on the stems and become covered with numerous conidiophores and spores; these are very easily dispersed by wind and water splashes. Contaminated agricultural machinery, tools and substrates spread the pathogen.

● **Conditions encouraging development:** Not much is known on this subject. We have observed an outbreak which occurred in a plastic tunnel, in the climatic conditions of a Chinese spring, in the Beijing region. It seems that this disease and verticillium wilt develop in similar conditions.

CONTROL

● **During cultivation:** There is no really effective method of control. Nevertheless, any plant showing early symptoms should be destroyed.

As in other fusarium wilt diseases, fungicides based on benomyl or carbendazim can be applied to the collar of the plant, either locally or via the irrigation system; this may check development of the disease.

Prunings should be collected and destroyed. When the crop-growing season is over it is essential to remove and burn the plants.

● **Following crop:** Where protected crops are affected the following measures are advisable:

● Sterilisation of the soil, preferably with steam;

● Disinfection of all equipment used in production and cultivation of the plants as well as the structures of the glasshouse, etc., with a 2% aqueous solution of formaldehyde or sodium hypochlorite.

The cucumber varieties at present on the market are all more or less susceptible. *Cucurbita ficifolia* can be used as a resistant rootstock.

Fusarium oxysporum f.sp. *melonis* (L. & C.) Snyder & Hansen

Causes: **Fusarium Wilt of Melon**

There are 4 strains: **strain 0, strain 1, strain 1–2, strain 2. (3 strains are found in France.)**

SYMPTOMS

Refer to 1, 73, 98, 205, 282, 283, 285, 286, 287, 288, 289, 358, 422.

MAIN CHARACTERISTICS

- **Frequency and economic importance:** This *Fusarium* is found in many countries in the Americas, Asia and Europe. It often causes great damage and it is not uncommon for all the plants in one crop to die. In France it is now one of the most serious diseases affecting melon plants.
- **Source:** The fungus can survive in soil on plant debris in the form of chlamydospores (resting spores), which have thick, resistant walls. It can live like a saprophyte, enabling it to colonise and survive on various types of organic matter. It penetrates the roots at the point where the latter emerge from the stem.
- **Spread:** Contaminated seeds may be responsible for transmission of the disease, although this is quite rare. The soilborne fungus is often spread via wind, water splashes, machinery and tools. When humidity is high, stem cankers become covered with numerous fungal fructifications and spores; these are easily dispersed and are an important source of the pathogen.
- **Conditions encouraging development:** There is a considerable difference between the *in vitro* optimum temperature of the fungus (28°–30°C) and the temperature at which it becomes particularly aggressive as a pathogen (18°–20°C). Outbreaks of the disease occur mainly in cold or late springs. When temperatures rise above 30°C, the disease regresses (but does not disappear) and infection reduces. Unlike other wilt diseases it nearly always occurs in cold soils, early in the season. It should be noted that the conditions favouring the appearance of this disease vary. Wilting is most marked when temperatures rise and the relative humidity decreases.

The nutrition influences the susceptibility of plants to this disease. A high nitrogen content will increase their susceptibility; a higher content of potassium and calcium will reduce it.

CONTROL

- **During cultivation:** There is no effective remedy at this stage. Diseased plants must be destroyed as soon as symptoms appear. The application of benomyl or carbendazim to the stem base does not have sufficient curative effect or give any real protection to the plants, even if done frequently. In the latter case, there is a danger of phytotoxicity and the cost is too great.

At the end of the crop season, it is absolutely essential to destroy the plants.
- **Following crop:** Crop rotation is ineffective unless melons can be excluded for at least 8 years.

Compost for propagation must be free from the pathogen; avoid placing your plants directly on the soil, especially if this has not been disinfected. Where premises are contaminated, the nursery should be disinfected (the soil and the structures) as well as equipment and the compost used. Do not store the compost in bulk outside the glasshouses and never mix with sand or soil taken from open ground.

Soil disinfection may be difficult to carry out, especially in the field, and the results can be disappointing because this *Fusarium* will rapidly recolonise treated soil. The best soil-sterilising fumigants[*] are chloropicrin, methyl bromide or a mixture of the two.

With protected crops it is advisable to disinfect the structures, the machinery and the irrigation system. A 3% formaldehyde solution or sodium hypochlorite can be used.

Avoid the use of fertiliser with an excessive nitrogen content, and ensure the correct balance with calcium and potassium.

The best solution is to cultivate varieties which are resistant to one or several strains (refer to the list of resistant varieties in Appendix III).

Four strains have been described, but only three are found in France: strains 0, 1 and 1–2.

IMPORTANT NOTE

[*] Approval of fungicides may vary from country to country and local labels should be consulted before use.

| | Strains | | | |
Resistance genes	0	1	2	1–2
None (susceptible control)	S	S	S	S
Fom-1	R	S	R	S
Fom-2	R	R	S	S
Fom 1 & Fom-2 and polygenic tolerance	R	R	R	R

Because of the widespread distribution of strain 1-2 in so many soils the genetic resistance may not be very effective. This applies, for example, to some protected cropping sites in the south-east of France, and field-grown crops in the south-west. The growers there have to rely on grafting: formerly, *Benincasa cerifera* was used; nowadays a hybrid between *Cucurbita moschata* and *Cucurbita maxima* is preferred. This hybrid is much hardier, can withstand cold spring soils, and is highly tolerant to lime and soil diseases.

Growers in the south-east should remember that their soil can be suppressive of this disease. The longevity of this suppressiveness should not be put at risk by fumigation of the soil.

The results of biological control with antagonistic *Fusarium* or with other micro-organisms have been disappointing, and in the short term do not appear to be very practical.

Fusarium oxysporum f.sp. niveum (E.F.Smith) Snyder & Hansen

Causes: **Fusarium wilt in Water Melon**

There are 3 strains:
strain 0: Common strain affects varieties without resistance
strain 1: Less virulent, attacks same resistant varieties
strain 2: Very virulent, attacks all resistant varieties now present in France.

SYMPTOMS

Refer to 301, 302, 303.

MAIN CHARACTERISTICS

• **Frequency and economic importance:** Worldwide, this is the most damaging disease to affect water melons, especially in the USA. Crop yield may be reduced by 75% in affected areas. It is found in many countries in the Mediterranean basin (Spain, Italy, Israel, North Africa, etc.).
• **Source:** When the fungus colonises a plot it can survive for more than a decade in the absence of water melons, living as a saprophyte on plant debris. For example, it can colonise dead water melon stems and the roots of various weeds. It persists for more than 2 years in water melon seeds. Infection is often via the lesions left by emerging lateral roots. The chlamydospores, which it produces from the mycelium or the macroconidia, are also an important potential source of the pathogen.
• **Spread:** The organism is very easily dispersed via seeds, contaminated soil dust which may be windborne, substrates and plants, irrigation water, tools, various types of agricultural machinery, etc. Necrotic areas covered with numerous fungal spores appear on the stems of diseased plants; the spores are dispersed mainly through water splashes.
• **Conditions encouraging development:** The fungus has an optimum development temperature of 26.5°C, but the wilt symptoms usually appear at higher temperatures in low relative humidity and bright light. During such conditions, the plant cannot compensate for the high evaporation loss because of the presence of the fungus in the vessels and also the plant's reaction to its presence. The plant vessels produce gum in an effort to stop the progress of the *Fusarium*, but this indirectly impedes the water flow. Soil containing a high proportion of organic matter favours the development of this disease. The disease is also more common when plants are attacked by gall nematodes which encourage fungal penetration into the roots.

CONTROL

• **During cultivation:** There is no effective control. It is nevertheless advisable to remove diseased plants as soon as the first symptoms appear.
Treating the stem base with benomyl or carbendazim does not give sufficient protection.
When the crop season is over, it is absolutely essential to destroy the plants, as the fungus will continue to propagate on them, which will aid its survival in the soil.
• **Following crop:** Crop rotation is not feasible as *Fusarium* persists for too long in the soil. Often it is wiser to rotate the growing area. Disinfection of the soil with fumigants (methyl bromide, etc.) is of doubtful value, because the fungus colonises the soil very rapidly and may contaminate a large number of plants. If this solution is adopted, it is effective for one season only; careful precautions should be taken to avoid early recontamination.
Soil improvement should be carried out in moderation. It is essential to use healthy seed from a source which is known to be free of the disease.

(Vascular fungi)

Resistant varieties are by far the most satisfactory and some are available: ask your supplier to get them for you. It should be noted that in some production areas (Israel), the strains have such high pathogenicity that there is no totally resistant variety. In Japan, water melon is normally grafted onto a resistant rootstock: *Lagenaria siceraria*.

The use of 'solarisation' (see Factfile 17) in the USA has succeeded in reducing the inoculum levels of the fungus in soil, with a consequent reduction in the incidence of the disease in heavily contaminated ground. Today this is not an option in France, it is not sufficiently effective.

Verticillium dahliae Kleb. *Verticillium albo-atrum* Reinke & Berth. (species found in more northerly areas)

Causes: **Verticillium wilt**

SYMPTOMS

Refer to 280, 290, 291, 292, 293, 298, 299, 300.

MAIN CHARACTERISTICS

- **Frequency and economic importance:** These pathogens have a wide host range and are found in many countries in temperate zones. They can attack many species of cucurbits. We have seen melon plants affected which have also been stressed by temperature and humidity.
- **Source:** These fungi can survive in soil over a long period owing to resistant structures (microsclerotia) and also because of the availability of numerous hosts, cultivated or wild plants (aubergines, tomatoes, black nightshade, amaranth, etc.).
- **Transmission:** Garden compost and tools are the source of plant contamination and transmission of these fungi. Conidia are easily dispersed by air currents in protected crops, as well as by water splashes and some soil insects.
- **Conditions encouraging development:** The disease is frequently most severe at relatively low temperatures (20°–23°C). Short days and low light levels increase the susceptibility of the plants to the disease. For the same reasons, plants which are set out too early in the year are more likely to be affected.

CONTROL

- **During cultivation:** There are few effective methods of controlling this disease. In the case of an early crop, field-grown or soil-less, benomyl or carbendazim (at the rate of 0.5–1 g active ingredient/plant) can be applied to each stem base as soon as symptoms are found. This treatment (repeated at 3-weekly intervals) should restrict fungal development in the plants; also ensure that temperatures are raised to a level that inhibits growth (above 25°C). In soil-less culture, the percentage of active ingredient should be reduced to avoid the risk of phytotoxicity (yellowing and necrosis of the periphery of the leaf lamina).

  All crop haulm and debris must be removed and destroyed at the end of the season.
- **Following crop:** It is inadvisable to plant melons in highly contaminated soils; particularly in south-east France, plant them instead on virgin land. When tunnels are infested, disinfection of the soil with a fumigant such as methyl bromide can be considered, although this may not be completely effective. The same applies to solarisation, the method using solar energy (see Factfile 17). In infected soil-less culture, the substrate can be changed or disinfected (see Appendix I).

  Today there is no resistant melon variety on the market, although there are varietal differences. Varieties of the 'Olive' or 'Canari' type appear to be more susceptible than Charentais cantaloupes.

(Vascular fungi)

VIRUSES

Factfile 26

Cucumber Green Mottle Mosaic Virus (CGMMV)

Causes: **Cucumber Mosaic Disease**

SYMPTOMS

Refer to **24, 62.**

MAIN CHARACTERISTICS

• **Frequency and economic importance:** This virus is commonly found in the areas of northern Europe where protected crops are grown, e.g. the Netherlands, UK, or in Japan. Strains of the virus have been found in water melons in Japan and in melons in India and Iran.

• **Transmission:** CGMMV is not transmitted by insect vectors. Infection occurs through the roots, when soil or recycled substrate contains infected plant debris. It can also be transmitted in contaminated irrigation water or nutrient solutions. The virus is very stable and is mechanically transmissible during pruning or harvesting, or simply by leaves rubbing against the infected leaves of the next plant.

The most effective and the most serious form of transmission is by seeds. In cucumbers, 8% of the seeds may be infected; this rate drops rapidly to 1% after a few months of storage.

• **Ecology:** Secondary spread of the virus is rapid by mechanical transmission during cultural operations. The danger of the virus spreading by means of contaminated irrigation water or plant debris should not be ignored.

CONTROL

• **During cultivation:** There is no cure; once a plant has been infected it remains so for the rest of its life. At the very beginning of an outbreak and especially if the symptoms appear when the seedlings are emerging, it is advisable to destroy the first diseased plants. However, it must be remembered that the symptoms appear only after an incubation period of 1–2 weeks, during which the virus has multiplied in the plant; it has probably already been transmitted to nearby plants so that by the time the first symptoms are observed, the epidemic may already be developing.

Disinfection of pruning and harvesting tools with 3% trisodium phosphate will reduce the mechanical transmission of the virus from plant to plant but will not prevent transmission through leaf contact.

If the virus is found in only one section of the farm, cultural operations such as pruning or harvesting should be halted on that section, to avoid introducing the virus to the remaining growing areas.

• **Following crop:** Because there is no known resistance to this virus, it is essential to sow healthy seeds. Contamination on the seeds is normally confined to their external surfaces and can be eliminated by dry heat treatment (70°C for 3 days), which does not adversely affect the seeds' germinating ability. The presence of the virus inside the seed can now be detected so that seed batches can be certified virus-free.

Soil contaminated with the virus can be disinfected with methyl bromide.

Squash Mosaic Virus (SqMV)

SYMPTOMS

Refer to **60, 411**.

MAIN CHARACTERISTICS

- **Frequency and economic importance:** This virus is found in many production areas (the Americas, Australia, Taiwan, south of the Mediterranean basin, etc.) where there are a large number of vectors. It causes considerable damage to courgettes and to some varieties of melon, particularly the Charentais type.
- **Transmission:** SqMV is transmitted by many species of coleoptera (among them *Acalymma trivittata*, *Epilachna chrysomelina* and *Diabrotica undecimpunctata*). The mandibles of the insects become contaminated through eating the leaves of diseased plants (a 5-minute feed is enough), and they then transmit the virus while feeding on healthy plants. The vectors can transmit the virus over a period of 1–3 weeks.

The most effective and the most serious method by which the virus is transmitted is by seeds. The market in seeds has resulted in the virus being introduced into many countries, such as France and Canada, where it did not previously exist. The proportion of seeds infected varies according to the strain and the host. It may be extremely high, up to 94%, as in the melon; in general, from 1–10% of a commercially produced contaminated batch can be affected. In courgettes the proportion is usually from 1–5%.

The virus is very stable and can also be transmitted during pruning or harvesting, or simply by leaves rubbing together.
- **Ecology:** SqMV is found in many areas where seeds are produced (Canada, Israel, Taiwan). Whenever it has been reported in France it has been introduced by contaminated batches of seeds, with the virus being subsequently spread mechanically. None of the known vectors are at present found in France; however, a species close

to one of them, *Epilachna argus*, is occasionally reported in the Midi. It may play a part in the epidemiology of the virus, in particular by transmitting it to reservoir plants such as *Ecballium elaterium*, on which vector insects develop. Other weeds, such as *Chenopodium album* and *Chenopodium murale*, may also transmit the virus via their seeds, enabling it to survive in the environment independently of a vector or susceptible crop.

CONTROL

- **During cultivation:** There is no effective cure for this pathogen; once a plant is infected, it remains so for life. Occasionally, however, the plant may recover. The young leaves which emerge after the worst symptoms have appeared may look almost normal. This makes diagnosis difficult; however, the virus is likely to remain highly active and such plants will continue to be potential sources of virus.

As soon as the outbreak occurs and particularly if symptoms appear as the seedlings are emerging, it is advisable to remove the first diseased plants, especially if the crop is under cover. It should be noted that symptoms are apparent only after an incubation period of from 1–2 weeks, during which time the virus has continued to multiply in the plant and may already have been mechanically transmitted to nearby plants. Therefore, by the time the first symptoms are noticed, the epidemic is likely to be well developed.

Disinfection of pruning or harvesting tools with a 5–10% solution of trisodium phosphate reduces the risk of mechanical transmission from plant to plant. If just one section of the farm is affected, all pruning or harvesting operations should be halted in that section to avoid introducing the virus into the other growing areas.
- **Following crop:** The most important safeguard is to sow virus-free seed. Methods of detecting the virus within the seed have now been perfected, and it should soon be possible to certify seed as virus-free.

(Viruses transmitted by seeds and contact)

239

(Viruses transmitted by fungi)

Melon Necrotic Spot Virus (MNSV) and other similar viruses

Causes: Necrotic Spot Disease of Melons
Necrotic Spots on leaves

SYMPTOMS

Refer to 138, 139, 140, 141, 142, 143, 183, 270.

MAIN CHARACTERISTICS

● **Frequency and economic importance:** This virus has affected early melon crops in southeast France for many years and it is quite common in early spring and autumn in protected crops (glasshouses, plastic tunnels), grown in soil and soil-less culture. The disease rarely affects field-grown crops. It has been reported in many countries, sometimes causing great damage. In Japan, losses due to the virus may amount to 10% of a region's total production.

● **Transmission:** MNSV is transmitted by the soil fungus *Olpidium radicale*, whose mobile zoospores swin in liquids particularly in the capillary water of soil particles. The zoospore may acquire the virus in the soil: the virus attaches to the zoospore's membrane and is transmitted when the zoospore penetrates the root of a healthy plant.

The virus's great stability and the presence of fungal resting spores mean that a contaminated soil will remain so for a very long time (several years).

Studies have shown that the virus is also very easily transmitted by mechanical means (in particular, during pruning work) and probably by leaf contact as well.

Melons are known to be infected via the seeds, where the contamination rate can be very high, from 1–22%. However, this is probably an indirect method: it does not occur when seeds are sown in a sterile substrate. On the other hand, the virus is frequently transmitted by seed when the soil contains the vector fungus. This is known as vector-assisted seedborne virus transmission.

In the USA, transmission of the virus by a coleopteran (*Diabrotica* sp., not known in the UK or France) has been reported.

● **Ecology:** The disease is present in many production areas in France and other countries with a variety of climates: Mediterranean (Greece, southern Spain in the Almeria region, the Midi in France). The virus infects only cucurbits; melons and cucumbers are the only species which might be infected in the wild.

The virus does not appear to rely on reservoir plants for its survival; it is more likely to remain in the soil or on contaminated seeds.

CONTROL

● **During cultivation:** There is no effective remedy: once a plant is infected, it will remain so for the rest of its life. The plant's susceptibility varies according to the environmental conditions. The disease appears to favour the periods in the year when temperatures and light intensity are low. At the very beginning of an outbreak it is advisable to destroy the first infected plants, particularly in soil-less systems. As the symptoms do not appear until some days after infection, the plant may already be a source of the virus for the fungus vector. In any case, if the inoculum is present in the soil or substrate, removing the plants will have no effect.

Treatment aimed at the fungus vector, especially in soil-less cultures, can be quite successful in limiting the spread of the disease. In the UK and the Netherlands, the addition of Agral at 20 mg/l in the nutrient solution has been shown to be effective for cucumbers grown on rock wool.

Disinfecting the pruning tools with trisodium phosphate or sodium hypochlorite will reduce the risk of mechanical transmission of the virus.

● **Following crop:** There are melon varieties which are resistant to this virus but at present these do not include any of the 'Charentais' type. Until such resistant varieties have been developed, there are a number of cultural measures which will effectively limit the development of the disease:

● Disinfect the soil (steam, methyl bromide, chloropicrin, etc.) before any susceptible crops are planted;
● Practise a regular crop rotation;
● Avoid re-using rock wool which has grown diseased plants, or disinfect it first;
● Add Agral to the nutrient solutions;
● Graft onto resistant plants such as gourds (*Cucurbita ficifolia*);
● Sterilise tools.

OTHER VIRUSES OF THE SAME TYPE

Several other viruses cause diseases in cucumbers similar to those caused by MNSV, characterised by necrotic symptoms on leaves and stems. All these viruses have a range of hosts vastly wider than MNSV and infect large numbers of species belonging to families other than the Cucurbitaceae.

Cucumber Necrosis Virus (CNV) has been reported only in North America. It is transmitted by *Olpidium radicale*.

Some strains of Tobacco Necrosis Virus (TNV) are found on cucumbers (144, 145, 146, 147) and cause damage in the Netherlands, in the UK, and in eastern France. They are transmitted by another species of *Olpidium, O. brassicae*. Two other viruses of this type are transmitted through the soil or by mechanical means but it has not yet been proved that they can be transmitted by fungi. Cucumber Leaf Spot Virus (CLSV) is transmitted by seeds to cucumbers; it has been reported in both south and north Europe. Cucumber Soilborne Virus (CSBV) has so far been reported only in Libya.

Today, these viruses can be distinguished only with the aid of very specific and generally serological laboratory tests.

Cucumber Mosaic Virus (CMV)

SYMPTOMS

Refer to 4, 5, 11, 13, 14, 19, 22, 45, 46, 47, 51, 53, 54, 55, 56, 61, 65, 84, 202, 218, 306, 389, 394, 398, 405, 412.

MAIN CHARACTERISTICS

• **Frequency and economic importance:** This is undoubtedly the most common virus infecting field-grown cucurbits worldwide; it is also common in protected crops. First reported at the beginning of this century in the USA, the virus is now found on every continent. Particularly active in summer and autumn, the damage it causes is all the greater because it infects young plants. Water melons are rarely affected.

• **Transmission:** CMV is transmitted by aphids, in a non-persistent manner. The vector is able to pick up the virus from infected plants, and transmit it to a healthy plant, during the course of extremely short puncture bites lasting 5–10 seconds ('test' punctures which enable the insect to tell whether it has landed on a host suitable for its development). As a general rule, the aphid can transmit the disease in this way over a period of 1 or 2 hours, but it rapidly loses this ability if it makes many feeding punctures. Many varieties of aphids are vectors, among them the melon aphid, *Aphis gossypii*, and the green aphid, *Myzus persicae.*

This highly efficient method of transmission means that the disease can spread through a crop without large numbers of aphids being visible on the plants.

Peppers grown in soil-less culture may become infected by contact with diseased root and plant debris, especially in a rock wool system; this has not yet been confirmed for cucurbits.

It does not appear to be transmitted by the seeds of cucumbers, melons or courgettes.

• **Ecology:** CMV is found in all production regions. Some strains thrive in hot climates (Mediterranean or tropical) while others are more active at lower temperatures (the temperate zones of northern Europe). It infects numerous different botanical species (700 have been recorded up to date), both annual and perennial, whether cultivated (tomatoes, peppers, lettuce, spinach, etc.) or growing wild (hogweed, shepherd's purse, starwort, black nightshade, etc.). The latter play a very important role as the winter hosts of the virus, acting as reservoir plants. In spring they then become a source of the virus, and sometimes also of the vector aphids which initiate the epidemics (see the epidemiology diagram on page 47).

CONTROL

• **During cultivation:** There is no effective cure against this virus; once a plant becomes diseased, it remains so for the rest of its life. At the beginning of an outbreak, it is advisable to remove and destroy all infected plants, particularly in protected crops. It should be noted that the symptoms appear only after an incubation period of 1–2 weeks, during which time the plant can be a virus source for the aphids. For this reason, by the time the first symptoms appear on the first plants the epidemic may already be underway.

Insecticides are useful for limiting the aphid population if the plants are badly infested. They are not so effective in controlling the development of virus epidemics as the vectors often arrive from outside the growing area and transmit the virus during very brief puncture bites; so brief, that the aphicide does not have time to work.

• **Following crop:** The first priority is to sow varieties resistant or tolerant to the virus. They are obtainable for some types of cucumbers and courgettes, and it is likely that others will be developed for the melon grower in the future. There are various cultural measures which can be taken to restrict or retard the development of epidemics; they are helpful on their own or when used in combination with resistant varieties:

● Protect the seedbeds and the young plants with very fine-mesh netting or non-woven plastic film such as Agryl P17. Unfortunately this type of protection cannot be continued after the flowers have appeared, or pollinating insects would be excluded;

● Keep the edges and hedges round the growing area cleared of weeds to eliminate some of the sources of viruses and/or vectors;

● Use transparent or opaque plastic surface mulches on the crops, particularly when field-grown, which will keep the aphids at bay;

● Avoid planting late crops close to earlier crops which may already be infected.

(Viruses transmitted by aphids)

Papaya Ring Spot Virus (PRSV) (formerly known as Water Melon Mosaic Virus 1 [WMV1])

Causes: **Water Melon Mosaic Disease**

SYMPTOMS

Refer to **18, 26, 33, 49, 59, 68, 104, 405.**

MAIN CHARACTERISTICS

● **Frequency and economic importance:** This virus is rarely found in mainland France, except in late crops. However, it is regularly isolated in the area around Nice, in Guadeloupe and Martinique, and in the southern parts of the Mediterranean basin, where it causes serious damage. There are many strains, and it is thought that certain viruses which have been treated as distinct types (such as **Water Melon Mosaic Virus type Morocco,** or **Zucchini Yellow Fleck Virus [ZYFV]**) **may in fact be specific strains of PRSV.**

● **Transmission:** PRSV is transmitted by aphids, in a non-persistent manner. The vector can pick up the virus from an infected plant, or transmit it to a healthy plant, during very brief feeding periods (e.g. 'test' punctures of up to one minute which enable the insect to find out whether the plant attacked is a suitable host for its development). The aphid can normally transmit the virus over a period of about an hour, but rapidly loses this capacity if it makes further 'test' or feeding punctures. Many aphid species transmit the virus, among them being the melon aphid, *Aphis gossypii*, and the green peach aphid, *Myzus persicae.*

The considerable efficiency of the transmission method means that the disease can spread throughout a crop without large numbers of aphids being visible.

From various observations, it seems possible that the virus can also be transmitted by mechanical means, during harvesting or pruning, or by leaf contact.

There has been no report of transmission through the seeds of cucumbers, melons or courgettes.

● **Ecology:** The virus is found mainly in tropical regions or countries bordering the Mediterranean; it appears from time to time in France, generally in the south-east. Chiefly it infects Cucurbitaceae species, but apparently not the two species found growing wild in France (bryony and *Ecballium elaterium*). For this reason the virus can survive only in regions where cucurbit crops are grown all the year round, e.g. the area around Nice. It is also possible that it could be introduced into France from countries to the south, either by flying aphids or from imported contaminated fruit.

CONTROL

● **During cultivation:** There is no effective cure; once a plant is infected it remains so for the rest of its life. At the very beginning of an outbreak it is advisable to remove the first infected plants, particularly in protected crops. However, it should be noted that symptoms appear only after an incubation period of from 1–2 weeks, during which time the plant may be a source of the virus for the aphids. Therefore, by the time the first symptoms are seen, the epidemic may already be developing.

Insecticidal treatments will help to restrict the aphid population, if the insects are present in great numbers on the plants. They are not, however, of much use in preventing the development of virus epidemics as the vectors often arrive from outside the growing area and transmit the virus during such brief feeding periods which do not allow time for the aphicide to work.

● **Following crop:** There are resistant varieties of melon and cucumber. The following measures will also restrict or retard the development of epidemics:

● Protect the seed beds and the young plants with very fine-mesh netting or with a non-woven material such as Agryl P17. Unfortunately, this type of protection cannot be continued once the flowers have developed because it would prevent access to pollinating insects.

• Use thermal transparent or opaque plastic surface mulches to keep aphids at bay particularly on field-grown crops;

• Avoid planting late crops close to earlier crops which may already be affected;

• If crops are grown throughout the year, keep a 1–2 week cucurbit free period in the area, to break the viral transmission cycle.

Water Melon Mosaic Virus 2 (WMV2)

SYMPTOMS

Refer to **19, 29, 32, 55, 57, 66, 390.**

MAIN CHARACTERISTICS

- **Frequency and economic importance:** This polyvirus is often found in France, especially on field-grown crops. It is one of the principal pathogens affecting Charentais-type melons, which are very susceptible to this disease. Symptoms on cucumbers are very inconspicuous and may not be noticed. Gourds show wide varietal differences. Standard varieties of courgettes are normally quite resistant although the yellow types, grown widely in the USA, are highly susceptible.
- **Transmission:** WMV2 is transmitted by aphids, in a non-persistent manner. The vector can pick up the virus from an infected plant, or transmit it to a healthy plant, during very brief feeding periods (e.g. 'test' punctures of up to a minute which enable the insect to find out whether the plant attacked is a suitable host for its development). The aphid can normally transmit the virus over a period of about an hour, but rapidly loses this ability if it makes further 'test' or feeding punctures. Many aphid species are virus vectors, among them being the melon aphid, *Aphis gossypii*, and the green peach aphid, *Myzus persicae*.

The considerable efficiency of transmission means that the disease can spread throughout a crop, without large numbers of aphids being visible on the plants.

There have been no reports of transmission through the seeds of cucumbers, melons or courgettes.

- **Ecology:** WMV2 is found in France in the main production areas, in particular the south-west (round Toulouse and Nérac) and the south-east (lower Rhône valley). The virus attacks some plants not belonging to Cucurbitaceae: cultivated (peas, some varieties of bean, spinach, etc.) and wild (groundsel, shepherd's purse, dead-nettles, fumitory). The latter are very important to the virus's survival during winter, acting as reservoir plants. In spring they are a source of the virus and also sometimes of the vector aphids which initiate the epidemics (see the epidemiology diagram on page 47).

The wide range of reservoir plants identified worldwide no doubt explains the ability of this virus to adapt to very different ecosystems, such as sub-desert regions (Californian and Arizona deserts), sub-tropical (central Florida), or temperate (Europe). It has never been found in the extreme south of Florida or in Guadeloupe.

CONTROL

- **During cultivation:** There is no effective cure; once a plant is infected it remains so for the rest of its life. At the very beginning of an outbreak it is advisable to remove the first affected plants, particularly for protected crops. However, it should be noted that symptoms appear only after an incubation period of from 1–2 weeks, during which time the plant may be a source of the virus for the aphids. Therefore, by the time the first symptoms develop, the epidemic may already be underway.

Insecticidal treatments may help to limit aphid populations, if these are present on the plants in large numbers. However, they are less effective in preventing the virus spreading as the vectors often arrive from outside the growing area and transmit the virus during brief feeding periods before the aphicide has time to work.

● **Following crop:** Today there are no resistant varieties, not even of melon. The following measures can restrict or retard the development of epidemics:

● Protect the seed beds and the young plants with very fine-mesh netting or with a non-woven material such as Agryl P17. Unfortunately, this type of protection cannot be continued after flowering, because it excludes pollinating insects;

● Remove all weeds from the growing area and the surrounding borders and hedges, to eliminate sources of the virus and/or the vectors;

● Use thermal transparent or opaque plastic surface mulches on the crops, particularly when these are field-grown, to keep the aphids at bay;

● Avoid planting late crops close to earlier crops, which may already be affected;

(Virus transmitted by aphids)

Zucchini Yellow Mosaic Virus (ZYMV)

SYMPTOMS

Refer to 6, 7, 8, 9, 12, 20, 21, 23, 27, 28, 29, 30, 31, 34, 35, 41, 48, 50, 52, 56, 58, 63, 67, 69, 70, 71, 72, 83, 219, 264, 307, 316, 381, 382, 391, 392, 395, 399, 400, 401, 402, 403, 404, 405.

MAIN CHARACTERISTICS

● **Frequency and economic importance:** This polyvirus was identified only about 10 years ago; it causes very serious damage in France and in the UK, as well as in other main production areas worldwide. There are many strains, differing in symptomology, host range, and vector transmissibility.

● **Transmission:** It is transmitted by aphids, in a non-persistent manner. The vector can pick up the virus from an infected plant, or transmit it to a healthy plant, during very brief feeding periods (e.g. 'test' punctures lasting up to a minute which enable the insect to find out whether the plant attacked is a suitable host for its development). The aphid can normally transmit the virus for a period of about an hour, but rapidly loses this ability if it makes further 'test' or feeding punctures. Many aphid species transmit the virus, among them being the melon aphid, *Aphis gossypii*, and the green peach aphid, *Myzus persicae*.

The considerable efficiency of transmission means that the disease can spread throughout a crop without large numbers of aphids being visible on the plants.

Some studies suggest that the virus is transmitted by mechanical means during harvesting or pruning operations, or by leaf contact. So far there is no evidence of transmission through cucumber or melon seeds. Recent results have indicated that there may be some transmission through courgette seeds, but this has not yet been proved.

● **Ecology:** Spasmodic outbreaks of the virus have occurred in France for some years, but it is now much commoner and appears much earlier. Normally it is found in the south-west, although in some years it may reach the Parisian area. The virus mainly affects cultivated cucurbits, but under laboratory conditions it will also infect wild plants belonging to other families (e.g. dead-nettles, ranunculus). The epidemiology of the virus is not yet fully understood. In particular, we do not yet know what the sources of the virus are in spring. Does it over-winter in reservoir plants? Does it enter France from more southerly countries, by flying aphids, or by imported contaminated fruit? Once the answers to these questions are known, more efficient control will be possible.

CONTROL

● **During cultivation:** There is no effective cure; once a plant is infected, it remains so for the rest of its life. At the very beginning of an outbreak it is advisable to remove the first affected plants, particularly in protected crops. However, it should be noted that symptoms appear only after an incubation period of from 1–2 weeks, during which time the plant may be a source of the virus for the aphids. Therefore, by the time the first symptoms are seen, the epidemic may already be developing.

Insecticidal treatments may help to limit aphid populations, if these are present on the plants in large numbers. However, they are less effective in preventing the virus transmission, because the vectors often arrive from outside the growing area and transmit the virus during such brief feeding periods that the aphicide has no time to work.

● **Following crop:** There are resistant varieties of cucumbers, and resistant melons and courgettes should be available in the near future. In the absence of these, the following measures will help to limit or retard development of the epidemic:

● Protect the seed beds and the young plants with very fine-mesh netting or with a non-woven material such as Agryl P17. Unfortunately, this type of protection cannot be continued once the flowers have appeared because it excludes pollinating insects.

● Use thermal transparent or opaque plastic surface mulches to keep aphids at bay, particularly on field-grown crops.

● Avoid planting late crops close to earlier crops, which may already be affected;

● There are attenuated strains of the virus which appear to be effective in protecting courgettes from the more virulent strains. If this cross protection treatment proves to be feasible on a large scale, it can profitably be used for late-season courgette crops.

(Viruses transmitted by whitefly)

Muskmelon Yellows Virus (MYV) and other viruses

Cause: **Cucurbitaceae Yellows**

SYMPTOMS

Refer to 43, 76, 77, 79, 80, 81, 82.

MAIN CHARACTERISTICS

• **Frequency and economic importance:** The virus is seen occasionally in France and in UK in protected crops. It has still not been completely characterised, but was first reported in the early 1980s very soon after the disastrous population explosion of the whitefly *Trialeurodes vaporariorum*, in protected crops. Spring crops seem to be the worst affected. Effect on yield has not yet been accurately evaluated.

• **Transmission:** MYV is transmitted by *Trialeurodes vaporariorum*, probably in a semi-persistent manner. The vector can pick up the virus from an infected plant, or transmit it to a healthy plant during long feeding periods which last up to an hour. This is apparently not a very efficient method of transmission as large numbers of the vectors are required to start an epidemic.

• **Ecology:** The virus is found chiefly in protected crops. As there is no reliable method of detection, we do not know whether it also infects field-grown crops where no symptoms are apparent. In addition, because of the difficulties connected with the study of this type of virus, there is no accurate information on the relationship of MYV and other viruses, also transmitted by *Trialeurodes vaporariorum*, reported in cucumbers in the Netherlands and Japan (known there as **Cucumber Yellows Virus** [CYV]), and especially in melon and cucumber crops in southern Spain. It is likely that all these viruses are strains of **Beet Pseudo-Yellows Virus (BPYV)**. This organism attacks a large range of botanical species, both cultivated (lettuce, beetroot, etc.) and wild (groundsel, shepherd's purse, etc.). It is known that weeds act as reservoir plants, ensuring survival of the virus during the winter or summer.

CONTROL

• **During cultivation:** There is no effective cure against this virus; once a plant has been infected it remains so for the rest of its life. The symptoms appear only after an incubation period of from 2–4 weeks, during which time the plant may be a source of the virus for the whitefly. For this reason the destruction of affected plants will probably have no effect on the progress of the epidemic.

On the other hand, insecticidal treatments or biological control using *Encarsia formosa* will reduce the whitefly population in a heavily infested crop and may slow down the development of virus epidemics. By reducing the population of vectors, the rate of disease development will be reduced.

• **Following crop:** There are as yet no varieties resistant or tolerant to this virus. It is, therefore, advisable to take the following steps to limit or control the progress of epidemics:

• Protect the seed-beds and the young plants with very fine-mesh netting or a non-woven material, such as Agryl P17. Unfortunately, this type of protection interferes with pollination and must not be used once flowering has begun;

• Keep the edges and hedges round the growing area and also the immediate vicinity of glasshouses, etc., free of weeds, as these may be a source of the virus and/or the vectors;

• Destroy the plants as soon as possible, once the crop is harvested: the whitefly can establish large populations on these plants;

• Avoid planting late crops close to earlier crops, which may already be affected.

OTHER VIRUSES CAUSING YELLOWS

Melon or Cucumber Yellows were formerly associated with deficiencies; today the symptom is due far more often to virus infection.

In France, the symptoms of yellows have recently been observed in field-grown crops and in plant-breeding glasshouses in summer or autumn. A virus belonging to the **luteovirus** group, transmitted via aphids in a persistent manner (the first of its type to be reported on the Cucurbitaceae), has been associated with this disease (75).

It appears quite frequently in cucumber, courgette and gourd crops, where it also causes the symptoms of yellows. In field-grown crops these symptoms are often masked by the mottling caused by other viruses.

In southern California, USA, a disease with similar symptoms to those of MYV has for some years been causing a great deal of damage to melon crops. The virus responsible, **Lettuce Infectious Yellows Virus (LIYV)** is transmitted by the whitefly *Bemisia tabaci* (78). The virus may also be present in the Middle East and there is a danger that it may reach France, if the vector spreads to any great extent.

There are of course other viruses which can cause, among other symptoms, the yellowing of old leaves: for example, **Cucumber Yellow Vein Virus (CYVV)** transmitted by *Bemisia tabaci*, which is a serious problem in the Middle East, or **Zucchini Yellow Mosaic Virus (ZYMV)**, transmitted by aphids.

(Viruses transmitted by whitefly)

Cucumber Toad Skin Virus (CTSV)

Causes: **Cucumber Toad Skin Disease**

SYMPTOMS

Refer to **15, 64, 85, 396, 406.**

MAIN CHARACTERISTICS

● **Frequency and economic importance:** This rhabdovirus regularly attacks spring crops of protected cucumbers, although normally only a small number of plants is affected. At present it is reported only in France, in the south-east and the region around Orléans. Overall crop damage is slight, but those plants which are affected cease to produce fruits.

● **Transmission:** CTSV has no known vector and is very difficult to transmit by means from cucumber to cucumber. It is, therefore, not likely to be spread between plants during cultivation operations such as pruning and harvesting. It is, however, advisable to destroy all affected plants; their productive life has, in any case, finished.

Cucumber Pale Fruit Viroid (CPFV)

Causes: **Pale Cucumber Fruit**

MAIN CHARACTERISTICS

● **Frequency and economic importance:** It is the only viroid reported worldwide which infects wild cucurbits. So far it has been reported only in the Netherlands where it infects protected spring crops of cucumbers; normally only a small number of plants is affected. Damage is moderate or insignificant, and the most clear-cut symptom is the pale green/yellow colour of the fruit; in addition, growth is reduced and the fruit looks pear-shaped. Sometimes the leaves may become rough and acquire a blueish colour and the flowers may be atrophied or remain almost closed. First reported in the early 1960s, CPFV has not spread over a wide area; in France and Belgium small numbers of yellow cucumbers have been observed but it was not possible to confirm the presence of CPFV.

● **Transmission:** CPFV is not transmitted by aphids or by soil-borne vectors, nor does it appear to be transmitted through the seeds. It may, however, occasionally be spread from one plant to another during cultural operations such as pruning or harvesting. It is, therefore, highly advisable to destroy all dubious plants.

(Viruses transmitted by whitefly)

APPENDICES

411 Young melon plant with discoloured veinal tissues on the first leaf—Squash Mosaic Virus (SqMV).

412 Numerous melons plants showing mottled leaves which appear abnormally serrated—Cucumber Mosaic Virus (CMV).

413 Melon leaf has a creased look and an abnormally serrated leaf edge—Phytotoxicity.

414 Abnormally filiform melon leaf resulting from incomplete leaf growth—Phytotoxicity.

APPENDIX I

DISEASES OF YOUNG PLANTS

(occurring after sowing or pricking out, or at planting out)

THE YOUNG PLANT AND ITS WEAKNESSES

Young cucurbit plants are subject to the same diseases as adult plants. There are, however, several major differences between the young and the adult plants:

• Their immaturity makes them far more susceptible to diseases and various stresses connected with temperature and nutrition;

• The parent plant's background: the plants often inherit genetic abnormalities from the parent plants. A more serious drawback is that when the parent plants have been affected by diseases transmissible via their seeds, the young plants will be infected very early on, and frequently more severely (411). If the proportion of contaminated seed is high, the crop's future will be jeopardised;

• The high density of plants in the nursery: this makes them particularly vulnerable, especially to virus diseases transmitted by aphids. It takes only a few virus-infected plants to result in rapid spread to large numbers of healthy ones (412). When diseases are transmitted by seeds and by contact, the operation of pricking-out will spread the disease further.

UNSPECIFIC SYMPTOMS

Young plants show symptoms which are similar to those seen on adult plants. We shall describe the most representative of these:

Abnormalities in the shape and colour of the cotyledons and the young leaves (413, 416) may follow, for example, virus infection or slight burning associated with chemical treatment. If the phytotoxicity is quite severe and occurs early on before the leaf has developed (in the meristems), its growth will be partially disymmetric but stunted (414) giving it an abnormal appearance. The conditions causing this deformity are not fully understood but may be connected with a dry period preceding the development of the young leaves, with the result that they become larger or multilobed.

The cotyledons of certain *varieties* may be considerably deformed; this abnormality, sometimes genetic in origin, is a consequence of delayed germination of the seeds. These are probably too old with a low viability and they certainly have a thicker seed coat. It is advisable to destroy these seed batches and their seedlings.

It can be difficult to identify a cause with symptoms, as we found when we examined those shown in **415**. The interveinal tissue of the plants growing on the peat is chlorotic. We rapidly discarded any idea of a parasitic cause and suspected that the trouble was nutritional; analyses, however, proved that this was not so. Curiously, these plants, pricked-out or not into another substrate, but placed in another environment, began to develop normally, while those left with the original grower continued to show symptoms. This is not an uncommon situation. It shows what an important influence environment has on plant growth, both directly on the plants and also indirectly via the substrate.

257

415 Chlorotic interveinal tissues on the first leaf of a melon plant; the periphery of the lamina is necrotic—Phytotoxicity.

416 Vein yellowing of the 2 cotyledons of a melon plant—Phytotoxicity.

417 Yellowing and necrosis at the edge of cotyledons of cucumber plants—Phytotoxicity.

418 Wilted and desiccated sections of young cucumber leaves—Phytotoxicity.

Carelessly applied herbicides, in the nursery or its vicinity, can cause a greater or lesser degree of damage, for example yellowing of the cotyledon veins and necrosis of the developing plants. Phytotoxicity may also occur following treatment of the substrate with fungicides (**417**) or an excess of added CO_2 to the nursery atmosphere (**418**).

Wilting and desiccation with or without yellowing is a frequent problem in nurseries. The condition is often known as 'damping-off' and is caused by many very different micro-organisms (see diagram on page 262). This term describes a non-specific symptom of a particular parasite which includes wilting and collapse of the above-ground parts of the plants (**419, 420**). Once the latter come in contact with the soil or substrate, they rapidly decompose and disappear: thus the term 'damping-off'. On closer observation, however, it will be seen that the plant is affected by yellowing, browning and rot concentrated on the roots, stem base, collar and tissues which are still soft and not yet woody (**421**). It is this decay which is the cause of damping-off of the above-ground parts of the plant, often to a very severe degree. Some vascular parasitic fungi, such as *Fusarium oxysporum* f.sp. *melonis* cause losses in the nursery (**422**), especially if the substrate or soil on which the plants are growing is heavily contaminated.

False damping-off has been reported in cucumbers. In these cases the stem is constricted at some distance above the ground. There may be several causes:
• A short period of water stress after germination;
• The substrate is too dry, or it contains too high a proportion of soluble salts.

Some nurseries have a problem with seed-bed flies which lay their eggs on the young plants; the development of the larvae in the hypocotyl causes an apparent damping-off. If a longitudinal cut is made through the hypocotyl, the feeding larvae are clearly visible.

All the insects and airborne parasitic micro-organisms affecting cucurbits are able to infect to a certain degree young plants both in nurseries and in the field, if ambient conditions are suitable. In France the most common and the most damaging is *Cladosporium cucumerinum*. This fungus invades cold, damp nurseries in spring, causing leaf spots and stem cankers (**423**).

GOOD QUALITY STOCK

Young plants are liable to be affected by many health problems, parasitic and non-parasitic. Starting with the principle that the quality of the future crop depends on the quality of the plants that produce it, it follows that the latter must be perfectly sound. If there is any problem relating to the plants, this must be accurately identified and resolved as soon as possible. Laboratory tests are often essential; **do not hesitate to call on the services of a specialist laboratory.**

To avoid the risk of any such problems arising, we advise you to carry out the following procedures.

419 Wilting and desiccation of the cotyledons and first leaves in young melon plants. The stem has become very spindly—*Pythium* sp.

419

420

420 When the outbreak comes later, the plants do not collapse but several leaves become yellow and desiccated—*Pythium* sp.

421

421 The roots have become yellowish and more/less brown in several places—*Pythium* sp.

It is essential to plan for a 'health break' of several days between two crops. During this period the surfaces and the structures of the glasshouses, etc., the equipment (plant containers, tools, shelving, pots, boxes, etc.) should all be disinfected to destroy any spores or other propagules. Several products can be used:
- A 0.3–0.5% hypochlorite solution (at 48° CHl) for soil-less culture, 4–7% for the structures;
- A 2–5% formalin solution (using a 38% formaldehyde commercial solution);
- A quaternary ammonium solution (*Bioroche* at 0.25 1/100 litres, *Sernet* at 0.4 1/100 l, *Hortiseptyl* POV at 1 1/100 l, Hortinet at 1 1/100 l).

After using some products it will be necessary to rinse carefully with water. High-pressure sprays should be used on the surfaces and structures of the glasshouses to ensure a thorough wetting.

Today formalin is the most commonly used disinfectant. It can also be used as a fumigant at the rate of 0.9 l of commercial solution per 100 m³. Potassium permanganate (at the rate of 360 g for the same quantity of formalin) is often added as an oxidising agent. During fumigation the temperature should be higher than 10°C with humidity between 50–80% (ensure that the inner walls are not wet). Leave the glasshouses closed for 24 hours and then ventilate them for at least 24 hours before planting. Formalin is sometimes applied as a spray at the rate of 3, 500 l/hectare.

During disinfection, **all necessary working precautions should be taken** (gloves, masks, etc., should be worn) as the products are to a varying degree poisonous (through contact or vapour) and corrosive.

SEEDS

It is advisable to use good quality seed. Where traditional cultivation methods are followed, growers should collect seed from perfectly healthy fruit only. Seed can safely be kept for 4–5 years, if storage conditions are good (low temperature, dry atmosphere).

If problems have been encountered in previous seasons, the seeds can be disinfected either with sodium hypochlorite, or with a fungicidal dressing such as thiram.

Stages in the development of a young plant and locality of principal sites of damage.

1 Inhibited germination of seed } Failure of seedling to
2 Affected hypocotyl   } emerge (pre-emergence damage)
3 Decay of main root.
4 Necrosis of roots and/or collar   } Damping-off (post-emergence damage)
5 Decay of roots and/or collar and/or stem base. }

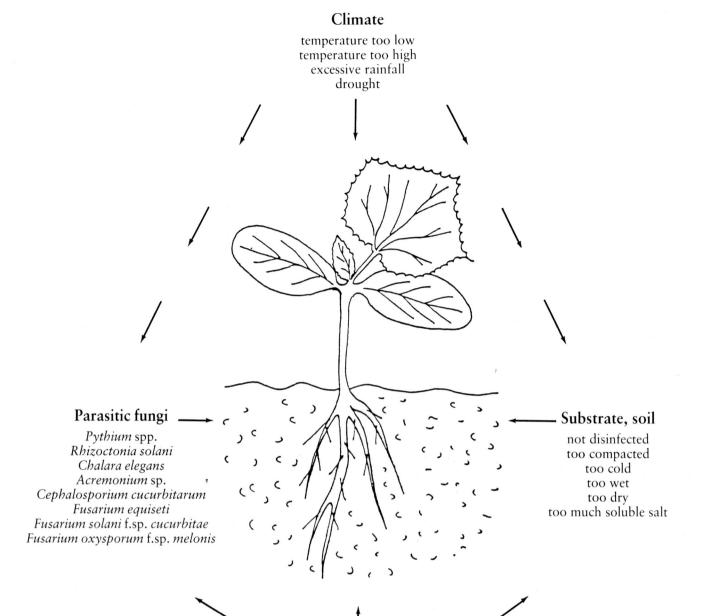

Climate

temperature too low
temperature too high
excessive rainfall
drought

Parasitic fungi ⟶

Pythium spp.
Rhizoctonia solani
Chalara elegans
Acremonium sp.
Cephalosporium cucurbitarum
Fusarium equiseti
Fusarium solani f.sp. *cucurbitae*
Fusarium oxysporum f.sp. *melonis*

⟵ **Substrate, soil**

not disinfected
too compacted
too cold
too wet
too dry
too much soluble salt

**Wrong measures taken
Bad cultural practices**

excess irrigation
lack of irrigation
enrichment with large quantities of organic matter
lack of day-time ventilation
lack of night-time heat
seeds sown too deep, plants planted too deeply

**Main causes, acting singly or in combination, of damping-off
and death of plantlets**

(* Most frequent in France and in the UK)

As a general rule, peat and loam are quite clean. To be certain, it might be safer to disinfect them with steam or a fumigant such as methyl bromide or formaldehyde, particularly if they have not been protected from contamination by soil during storage.

They should never be mixed with sand or loam of doubtful origin, i.e. material which has not been shown to be free of problems. Loam is very often contaminated, particularly by the Pythiaceae, and this is commonly the basic cause of many failures in the nursery.

When the soil blocks are being prepared, care must be taken not to compact them excessively—the block presser should be correctly set—otherwise there is less oxygen available for the seedlings and the block takes longer to warm up. They should not be over-watered.

An anti-Pythiacious fungicide such as propamocarb HCl can be mixed into the substrate before the blocks are made. When these are ready, they can be wetted with a fungicidal solution combining propamocarb HCl and liquid Cryptonol (1 ml of each product/litre of water). Etridiazole can be used in the same way, by mixing it into the substrate or by soaking the blocks. The addition of quintozene to the substrate is sometimes recommended to suppress *Rhizoctonia solani*.

PRODUCTION METHODS AND CONDITIONS

It is a mistake to sow the seeds or plant the young plants too deeply, as this makes them far more prone to damping-off.

Never place the growing blocks directly on the floor of the nursery, particularly if it has not been disinfected. It is better to place them in rows on shelves (disinfected) or on plastic sheets laid over the floor. These plastic sheets can be covered with a layer of peat, which should be renewed frequently. The plastic sheet should be replaced annually or whenever it is torn. It should be noted that raising several batches of seedlings in the same place may lead to an accumulation of surface salts which form a skin. The salts will then rise in the blocks and poison the plants, stopping their growth and burning the roots. The soil can be flooded or the plants can be isolated on plastic sheets. Once the blocks are in place, ensure that the climatic conditions in the nursery are suitable for plant development. Attempts to save energy, especially at night, very often cause considerable damage. Temperatures must be regulated to maintain warm conditions (minimum of 18°–20°C); avoid excess humidity in the substrate or atmosphere. On the other hand the blocks must not be allowed to dry out too much, or there is a danger of excess salinity; moreover, it becomes extremely difficult to restore the moisture content of the block.

Light and plant density both influence plant quality. Spindly plants are much more susceptible to disease. Plants should, therefore, be pricked out and replanted at the right moment; it is a mistake to delay the operation for too long.

All abnormal or unhealthy-looking plants should be immediately removed.

The control of virus diseases in the nursery is particularly important. The following elementary measures can be taken:
● Destroy any weeds in the nursery and keep the borders weeded;
● Watch out for aphids and apply prophylactic anti-aphid treatments;
● Avoid raising plants at a time when a crop is still in production or in an area close to diseased plants. A 'health break' and insecticidal treatments applied during this period can only be beneficial.

422 Yellowing and wilting of the cotyledons and leaves of melon plants; the stems are necrotic at the surface and the vessels are brown—*Fusarium oxysporum* f.sp. *melonis*.

423

423 Young gourd (left) and melon (right) plants attacked by *Cladosporium cucumerinum*. The more developed gourd plants show only a few leaf spots. The melons, at the stage where the two cotyledons have opened out, are badly affected; cankers can be seen on the epicotyl and heads have been lost from some plants—*Cladosporium cucumerinum*.

DISINFECTION OF SOIL AND SUBSTRATES

Soil preparation

Soil disinfection should be carried out in the right conditions to be effective: the soil, therefore, needs to be prepared. Several steps are required:
• Soil cleaning: remove as many roots as possible left from the previous crop;
• Tilling the soil: it should be subsoiled and ploughed. The top layer should be harrowed until it is as fine as a seed-bed, and properly levelled;
• Humidification of the soil: it should always be kept moist for as far down as possible, in particular just before disinfection;
• the soil temperature should be at least +15°C.

Suitable disinfectants

Steam can be used though few farms are equipped for this. Methyl bromide and/or chloropicrin (in countries where these are authorised) are the most effective products. **Methyl bromide** leaves residues in the soil which can harm crops grown in rotation, such as salad crops. In addition, a specialist firm has to be employed.

Dazomet or **metham sodium** are perfectly effective, if they are applied very carefully:

- Dazomet: 500–700 kg active ingredient/ha.
  - spread the product early in the morning;
  - dig it in quickly with a rotovator or rotary spade;
  - water immediately to encourage a crust to form;
  - do a cress test before planting.

- Metham sodium: 1, 200–1, 500 l of commercial product/ha (minimum soil temperature +10°C).
  - feed the product into the spray irrigation system; the measuring pump should be set to deliver the correct dosage/hectare;
  - drench at the rate of 30 to 40m³/ha;
  - keep the soil moist for 8–10 days;
  - till the surface soil so that the product is broken down;
  - do a cress test before planting.

Soil disinfection by sunlight (solarisation) could be considered, in particular when there are superficial parasitic fungi in the soil. Results have been striking in some Mediterranean countries. The method consists of covering the soil to be disinfected (well prepared and wetted) with a polyethylene sheet, 25–40 microns thick, and keeping it in place for at least a month during the sunniest period of the year (July and August).

Disinfection of soil-less substrates

Today very little is known about the disinfection of these substrates. Steam treatment is effective if its distribution can be adapted to the existing systems.

It is recommended that the substrate should be allowed to dry out as much as possible before treatment, in particular when dealing with rock wool blocks which can conveniently be packed into a forcing frame, so that the steam can circulate freely between the blocks. Where peat-based substrates are concerned, excessive drying-out causes 'mini-pockets' of air to form, which hinder the free circulation of steam.

The blocks should be saturated to 1/3 of their capacity and the temperature maintained at 100°C for at least 30 minutes.

Metham sodium cannot be used for organically based substrates, but is suitable for any material substrate (sand, Perlite, pozzolano). The substrate, initially dry, is repeatedly impregnated with the solution and then thoroughly washed.

Methyl bromide can be used; the method will depend on the type of applicator.

Formalin has been successfully used to disinfect pozzolano as well as peat; the substrate is soaked with a 3% commercial formalin solution.

An ideal solution would be insect-proof nursery premises. This can be achieved by blocking the openings with very fine-mesh netting. Covers made from non-woven materials, such as Agryl P17, are a good alternative.

Foot baths should be provided at the entrances to nurseries and glasshouses, and traffic should be restricted as far as possible.

Plants should be set into well-drained soil which is neither too cold nor too wet.

424

425

427

426

APPENDIX II

DAMAGE CAUSED BY THE MAIN PESTS AFFECTING CUCURBITS—PLANT PARASITES

MITES

Tetranychus urticae (red spider mite, yellow spider)
Tetranychus cinnabarinus
Tyrophagus similis
Tyrophagus longior

	Tetranychus urticae	*Tyrophagus similis* *Tyrophagus longior*
Description of pest	Globular mite, greenish/yellow, 0.3–0.5 mm long (**426**). Eggs spherical, translucent, 0.1 mm diameter.	Mite, white to fawn, 0.3–0.4 mm long.
Damage caused	Stops leaf growth. Small yellow dots on leaves (**424, 425**) which become chlorotic and faded; numerous silky webs (**426, 427**).	Distortion of young leaves (**428, 429**). Small pits on leaves which become larger as the leaves grew. Death of apical buds.

Tetranychus cinnabarinus looks similar to *T. urticae*, but is bright crimson.

428

429

430

432

43[

43[

APHIDS AND WHITEFLIES

Aphis gossypii
Macrosiphum euphorbiae
Aulacorthun solani...
Trialeurodes vaporariorum (whitefly)
Bemisia tabaci

Aphids	Whitefly
Description of pest	
Inactive insects 1.5–2.5 mm long, colour varies according to species (green, black, pink, etc.) (**430**). Apterous (wingless) and winged varieties.	Small white winged insect 1.2–1.5 mm long (**433**). Flattened larvae, oval and ciliate.
Damage	
Growth check. Distortion, shrivelling of the leaves. Produce honeydew covered with fumagine (**431, 432**).	Produce honeydew covered with sooty mould (**431, 432**).

Efficient virus vectors

How to distinguish between the most common whiteflies (appearance of pupae)

Trialeurodes vaporariorum   *Bemisia tabaci*

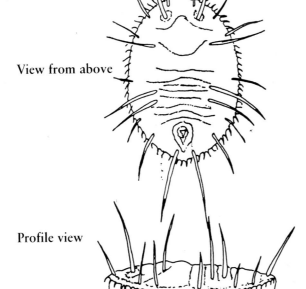
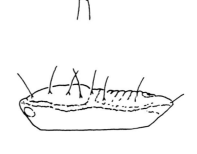

View from above

Profile view

(after Tong-Xian Lui)

434

43

43

436

Vector of the Tomato Spotted Wilt Virus (TSWV) on water melons in Japan; it also infects many other market garden species.

THRIPS

Thrips tabaci
Frankliniella occidentalis
Thrips palmi (in the West Indies, Japan . . .)

Thrips tabaci	Frankliniella occidentalis
Description of pest	
Small insect about 1 mm long, pale yellow-brown/black. Minute eggs inserted into the leaves.	Small insect about 0.9–1.2 mm, pale yellow/dark brown (437). Tiny eggs inserted into the flowers, leaves and fruit.
Damage	
Small silvery spots dotted with shiny black specks; rapidly becoming necrotic (434).	Discoloured, silvery patches on leaves (435) which become necrotic (436). Greyish patches on petals (437). Small greyish pits (438); their proliferation makes the fruit corky and dull (439, 440).

438

439

440

441

442

MOTHS

Chrysodeïxis chalcites...

Description of pest

Adult: butterfly with 3.5–4 cm wingspan, mottled brown with very pale gold spots. Caterpillar: 4 cm long, pale almond green (**442**) with fine whitish lines on the back.

Damage

Leaf perforations (**441**)
Fruit is superficially browsed (**443**).

443

LEAF MINERS

Liriomyza bryoniae
Liriomyza strigata
Liriomyza trifoli

Description of pest

Adult: very mobile 'fly', 2 mm long, yellow and black.
Larva: 1 mm long, yellow, burrows into the tissue of a leaf.
Pupa: barrel-shaped.

Damage

Tiny yellowish dots (feeding punctures) and numerous winding tunnels on the leaves (**444, 445**); later the leaves become desiccated.

445

DODDER

Many species of the Cuscutaceae have been reported on the Cucurbitaceae; affected plants become less vigorous and may die if this parasitic plant is allowed to develop. It is characterised by its suckers attached to young stems, leafless twining stems and stems and small flowers in clusters (**446**). Dodder may act as a vector, transmitting viruses from one plant to another.

446

447 Smooth-skinned Charentais cantaloupe sectioned by ridges; orange flesh.

448 Semi-netted melon, ridged, with orange flesh.

449 Elongated, yellow Canary melon with white or green flesh.

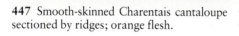

450 Round netted Galia with green flesh.

451 Fully netted American cantaloupe with orange flesh.

452 Late Spanish melon which keeps well; white flesh.

Various types of melon

APPENDIX III

SOME DATA ON CULTIVATED CUCURBITS AND RESISTANT VARIETIES

BOTANY OF THE CUCURBITACEAE

The Cucurbitaceae family includes several species of considerable economic importance in many countries on every continent, among them being the water melon, cucumber, melon and gourd (squash). Others are cultivated on a small scale and are of only very local interest: bottle gourd, vegetable sponge (loofah), bitter cucumber, chayote, etc. In France and the USA the main species grown for sale are the melon, cucumber and gourd, while in the UK the squash is gaining in popularity.

All these plants originated in hot countries (the Mediterranean or inter-tropical regions) and need fairly high temperatures. Some, belonging to the genera *Cucurbita* (*C. ficifolia*, *C. foetidissima*, etc.), *Cucumis* (*C. heptadactylus*, etc.), *Sechium*, are perennials in their native country and cannot survive the winter frosts in France. Melons, and in particular water melons, need high temperatures and will grow well in very dry conditions, while other species, for example *Benincasa*, are adapted to wet tropical conditions.

Only two species are native to France: bryony (*Bryonia cretica* subsp. *dioica*) that grows in hedges and is perennial due to its tuberous roots, and the exploding cucumber or donkey's cucumber (*Ecballium elaterium*), which is found in the hottest regions.

FLOWER BIOLOGY

The Cucurbitaceae are in general monoecious plants, i.e. male and female flowers are found on the same plant. Pollination is assured by insects, mainly bees but also by bumblebees. There are variations on the monoecious sexual type:

- Many melon varieties are andromonoecious and produce male and hermaphrodite flowers on the same plant; this is the case in particular with Charentais melons (before the plant breeders introduced the monoecious character), the netted American cantaloupes, the winter olives, the Canaries, etc.
- Some cucumber varieties are gynoecious and produce only female flowers. To make sure that fruit develops, there are two courses: either grow them together with other plants which have male flowers (monoecious pollinators)—this is the method used for growing some varieties of gherkin—or obtain parthenocarpic plants which do not have to be pollinated in order to bear fruit (glasshouse cucumbers, gherkins).
- Species belonging to the genus *Telfairia* and ecotypes of *Ecballium elaterium*, are dioecious in some regions.

It should be noted that the flower biology may, at least temporarily, be changed by spraying on growth substances. The gibberellins and silver nitrate will cause gynoecious plants to produce male flowers (thus enabling seeds to be produced through self pollination); ethrel suppresses the male flowers on monoecious plants (which does away with the necessity for the manual removal of the male flowers when producing F1 hybrids of monoecious melons or courgettes).

453 Spiny cucumbers ('Picking' variety) are picked very young for preparation as gherkins.

454 'Marketer' spiny cucumber ('Slicer' variety).

455 'Bet-Alpha' cucumber, widely grown in the Middle East.

456 Cylindrical glasshouse cucumber ('long' variety), smooth-skinned and parthenocarpic.

Some types of cucumber being cultivated today

As the Cucurbitaceae are allogamous (cross-pollinated by insects) their natural shapes and the native varieties which growers allow to inter-pollinate freely represent the background population. The basic material is relatively heterogeneous and heterozygous; therefore, to standardise it, pure homozygous lines or heterozygous F1 hybrids have been created and grown.

By definition, two plants belonging to different species cannot fertilise each other to produce hybrids (at least not under practical conditions). Thus a melon cannot be crossed with a cucumber to produce a hybrid, nor can a courgette with a pumpkin, still less a melon with a courgette. But two varieties of the same species can produce a fertile hybrid. The courgette and the decorative false colocynths can produce hybrids, as can two types of melon such as Charentais and Canary.

PRINCIPAL GENERA

Genus *Cucumis*

Two species of this genus are cultivated and have great economic importance: the melon (*C. melo*) and the cucumber (*C. sativus*). Others are of more marginal importance, such as *C. anguria*, produced for its small fruit which are eaten like cucumbers or gherkins in some tropical countries (the West Indies, etc.), or *C. metuliferus*, sold under the name Kiwano, horned cucumber or metulon.

The **melon** probably originated in inter-tropical Africa (all the wild species having $n = 12$ as the basic number of chromosomes are African). All the cultivated forms are diploid ($2n = 24$); they include many different types which are used in a number of ways:
- Eaten immature and raw, in salads, often confused with a cucumber in North Africa (fakous, etc.);
- Harvested very young and pickled like gherkins ('common' variety in the Far East);
- Harvested unripe and cooked in soups or in the oven;
- Picked when ripe (cantaloupes, winter olive, Canary, etc.) and eaten as they are because they are sweet.

The fruits also vary in shape (round, oval, pear-shaped, elongated, etc.) (**447–452**); thickness; colour and ornamentation of the rind (sides more/less marked, rind netted or smooth, etc.); colour of the flesh (green, white or orange); sweetness; weight (from 50 g to more than 10 kg); colour and size of the seeds, etc.

The physiology of ripening may take two forms:
- Varieties which go through a physiological change, i.e. an increase in respiration and ethylene production during the few days preceding maturity of the fruit (peduncular dehiscence). Fruit from these varieties does not keep well, an example being the Charentais melon.
- Varieties without this physiological change, whose fruit keeps as well as those above (early Canary or Tendral varieties) or longer (late Tendrals will keep for several months).

The **cucumber** is a native of India and is distinguished from all other species of *Cucumis* by the number of its chromosomes ($2n = 14$). It was grown and enjoyed by the Romans. There are variations in skin colour (green and white), the hairs and spines on the skin, and the shape (globular or long). The main cultivated varieties are differentiated chiefly by the shape of the fruit:
- Small spiny fruit harvested very young for making gherkins (**453**);
- Short fruit, spiny and cylindrical (Marketer) (**454**);
- Short fruit, smooth and shiny (Bet Alpha) (**455**);
- Long fruit, cylindrical and parthenocarpic, grown under glass (**456**).

457

458

459

457 Green courgette.

458 Yellow courgette.

459 Striped courgette.

461

460

462

460 Grey courgette.

461 Round courgette.

Cucurbita pepo fruit eaten in an immature state.

462 Angular peduncle of *Cucurbita pepo* (courgette) with awns extending onto the fruit.

Genus *Cucurbita*

All species of gourds originated in the Americas, from the south of the USA down to South America. They are all diploid (2 n = 40) and are characterised by their large, orange-red flowers. Three species are now grown in Europe, and a fourth is used as rootstock (*C. ficifolia*). The distinguishing properties of these four species are summarised in the table below.

They are generally grown for their fruit which is eaten when immature (courgettes) or when mature (pumpkins), but they are also grown for their decorative fruit which keeps for a varying length of time (colocynths, turk's cap pumpkins), their flowers eaten traditionally in the Mediterranean region, and their seeds which are eaten toasted or for their medicinal properties. The shape of the fruit varies widely (round, flattened, cylindrical, elliptic, discoid, pear-shaped) as does its, colour (green, grey, red, yellow) which may be an all-over coloration or in spots, stripes, etc.

	C. pepo	*C. moschata* *C. mixta*	*C. maxima*	*C. ficifolia*
Seed colour	White	White	White	**Black**
Leaf shape	Triangular very lobed	Triangular scarcely lobed	Rounded scarcely lobed	Rounded, lobed (figleaf)
Macula present	Yes	Yes	Few	Yes
Type of hairiness	Spiny (knobs)	Hairy	Hairy	Hairy
Appearance of sepals	Narrow	Leaf-like	Narrow	Leaf-like
Shape of section through fruit peduncle	Angular (5 angles)	Angular (5 angles)	**Circular (corky)**	Angular (5 angles)
Attachment of peduncle to fruit	**Continuous**	**Widened**	Retracted	Widened
Appearance of fruit epidermis	Shiny	Matt (bloom)	Shiny	

The descriptions in **bold** are the most characteristic features of the species. The others vary within a species from one variety to another. *C. ficifolia* can be recognised by its black seeds and oval fruit spotted green and white. *C. maxima* is characterised by the fruit's cylindrical peduncle which may be rather corky and irregular when the fruit is mature. *C. pepo* and *C. moschata* have a five-angled fruit peduncle. These angles look as though they extend onto the fruit in *C. pepo* (**462**) whereas in *C. moschata* the peduncle appears to have shrunk and become dissociated (**467**). The fruit of *C. pepo* is normally shiny while that of *C. moschata* is duller with a bloom.

463 Example of the various types of *C. pepo* (pumpkin, courgette, summer squash, colocynths, etc.).

279

464 Mature pumpkin, eaten mostly in the south-west of France.

465 Cylindrical peduncle of the fruit of *Cucurbita maxima* (pumpkin).

466 A ripe Provençal musk gourd fruit (*Cucurbita moschata*), eaten frequently during winter in the south-east.

467 Angular peduncle of *Cucurbita moschata* becoming retracted and dissociated from the fruit.

Cucurbita fruit eaten when mature

- *C. pepo* is the species requiring least heat. The most common varieties are courgettes (elongated or round, dark green to light grey in colour) (**457–461**), summer squashes, pumpkins (large fruit of inferior quality, normally used for animal feed), decorative colocynths (small fruit with a hard rind, long-keeping, very colourful and ornamental) (**468, 469**), 'vegetable spaghetti' (fibrous flesh resembling spaghetti).
- *C. maxima* can also tolerate lower temperatures. It is characterised by its peduncle which is circular in shape (**465**). This includes the winter squashes whose size and colour vary widely (Rouge vif d'Etampes, Potimaron). Some have a swelling where the pistils are attached, which is decorative (turk's cap pumpkin or turk's cap squash).
- *C. moschata* and *C. mixta* require more heat and are cultivated mostly in the south of France. The species includes the 'Provençal musk gourd' (**466**), the 'Pleine de Naples', picked when mature and eaten in autumn and winter, and also the 'long courgette' grown around Nice and harvested young with its flower.
- *C. ficifolia* is the only gourd species to have black seeds. In its original home (Central America) it is perennial and used as a rootstock for cucumbers to give resistance to soil diseases and pests (*Fusarium*, nematodes, etc.). It does not require a hot climate because it originated in mountain areas.

468 Young fruit of the summer squash, another member of the genus *Cucurbita* (*C. pepo*).

468

469

469 A decorative colocynth fruit, belonging to the same species as the courgette (*Cucurbita pepo*).

281

470

470 Water melon fruit (*Citrullus lanatus*), the most widely grown cucurbit in the world.

471

471 Young *Benincasa* fruit, eaten in south-east Asia and China; also used as rootstock for the melon in France.

472

472 Fruit of *Momordica charantia* or 'bitter cucumber' widely cultivated in south-east Asia.

Genus *Citrullus*

The **water melon** (*Citrullus lanatus*), originating in Africa, is the most important cucurbit worldwide. It is commonly known as 'water melon' in English, 'Wassermelone' in German, 'melon d'eau' in French, and locally in Provence as 'citre', 'gigerine' or 'méreville' (**470**). It is characterised by its highly denticulate leaves and its seeds which are scattered throughout the flesh of the fruit. The colour of the flesh can be red, yellow or greenish-white. The seeds are black, red, brown, etc. The rind may be one colour all over, marbled, striped or ridged and the fruit is round or elongated.

Water melons need high temperatures and are grown in very dry, sub-desert regions. The types most frequently found in France are 'Sugar Baby' with small, spherical fruit (3–4 kg), uniformly dark green, and 'Crimson Sweet', whose light green fruit is streaked with white. Green-fleshed types are grown in Provence for use in the confectionery and preserves business.

The seeds and oil extracted from them are also eaten. The similar species which grows wild, *C. colocynthis* (the true colocynth) is extremely bitter but it is harvested for its seeds.

Other genera

Members of the genus ***Benincasa*** are grown in wet tropical regions, chiefly in Asia. Their fruit (**471**) is harvested when immature and eaten cooked like courgettes. They are either smooth or very hairy and when mature can weigh from 20–30 kg. They are sometimes used as a rootstock for melons to give resistance against certain soil pests and diseases (*Fusarium*, nematodes).

The fruits of ***Luffa*** must also be harvested when very immature. They then later develop the network of fibres which will become a **loofah** (**vegetable sponge**). Two species are grown: ***Luffa cylindrica***, with fruits which are circular in section, and ***Luffa acutangula***, star-shaped in section.

The fruit in the **bottle gourd** (***Lagenaria***) is also eaten cooked when harvested quite young. The species is characterised by its large white flowers (whereas nearly all the other cucurbits have yellow flowers). For thousands of years it has been used as a receptacle in many different civilisations, both in Africa and tropical America. Pliny the Elder recommended its use as a buoy, when teaching children how to swim. He was mistaken when he wrote that seeds taken from the narrow part would result in plants producing cylindrical fruit, whereas seeds taken from the swollen part would give plants having round fruit!

Sechium edule is one of the rare members of the Cucurbitaceae to have only one seed per fruit. The **Chayote** (or **Christophine** or **Chouchou**) has a pear-shaped fruit coloured white to dark green, smooth or spiny. The plant is able to grow as a perennial on the French Mediterranean coast or in very sheltered spots during winter. The fruit is eaten either raw or cooked.

Momordica charantia is widely cultivated in south-east Asia (**472**), where it is known as **bitter cucumber**. The fruit is elongated and covered with irregular knobs and protuberances and is harvested young. When mature the fruit is orange and splits unevenly so that the seeds appear to be wrapped in the vermilion red pulp.

Trichosanthes anguina is another cucurbit with white flowers. The petals are very finely fringed. The pale grey fruit is very elongated (hence its name of 'vegetable snake') and is usually harvested before reaching maturity.

PRINCIPAL USES

Cucurbits are generally grown for their fruits, some of which may be used as animal food (pumpkins, green-fleshed water melons) but mostly for human consumption. The fruits may be harvested while still immature (cucumbers, courgettes, summer squashes, loofahs, bitter cucumbers, bottle gourds) or ripe, and eaten raw or cooked. The ripe fruits vary in the length of time they can be kept: a few days at ambient temperature for a Charentais melon, some months for a winter olive type melon or a Provençal musk gourd.

Cucurbits are grown also for their seeds which are normally very rich in high-quality oils (courgettes, water melons, telfairia). The shoots and leaves of some species are cooked and eaten, as are the tuberous roots of others.

The fruit of the vegetable sponge when mature develops a very strong fibrous network which allows it to be used as a sponge (hence its name). The fruit of the bottle gourd becomes extremely hard and is used as a receptacle or musical instrument in many tropical countries. The false colocynths are fairly hard and will keep for some time; their colour, shape and rind markings are very decorative.

The medicinal properties of this family are by no means negligible; for example, gourd seeds contain substances which facilitate the detachment of taeniae (tapeworms) from the intestinal wall, and also some contain substances which act as a decongestant on the prostate.

The Cucurbitaceae may also contain very bitter substances, the cucurbitacins, which can be poisonous. They are normally eliminated by selective breeding and are found only occasionally in certain parts of the plant (for example, all Charentais melons have bitter roots).

RESISTANCE OF CUCURBITS TO DISEASE

Resistance genes in the Cucurbitaceae were isolated some time ago, and this knowledge has been used to control disease; for example, the resistance to *Cladosporium cucumerinum* in cucumbers, or the resistance to *Sphaerotheca fuliginea* strain 1 in melons has been exploited for over half a century.

However, there is a limit to genetic resistance. On the one hand, the level of resistance may be insufficient: for example, it has never been possible to create a melon variety which is really resistant to Water Melon Mosaic Virus type 2. On the other hand, the stability or durability of certain resistances may not be satisfactory. A pathogen, like every other living thing, has its own variability and it may be capable of developing strains which can overcome or avoid certain forms of resistance, which then become ineffective against the pathogen. A good example is that of melons and *Fusarium oxysporum* f.sp. *melonis* where four strains are recognised. Finally, some types of resistance may be closely linked with deleterious characters which make their use difficult or even impossible. Thus very high levels of resistance to some mildew strains are associated with a necrotic symptom in melons (crown blight) or with chlorosis in cucumbers, during periods of short days and low light levels.

The heredity of resistance can be quite simple: one or two genes may control the resistance reaction. In such situations the breeder can introduce the genes relatively easily into new varieties. High resistance levels are often obtained (*Cladosporium cucumerinum* in cucumbers, *Fusarium oxysporum* f.sp. *melonis* strains 0 and 1 in melons, etc.) and the cultivated varieties should have responses of the all-or-nothing type (i.e. they are either susceptible or resistant). These genes are not always totally dominant; if the marketed variety is an F1 hybrid having only one parent with the resistant gene, the variety will be slightly less resistant than the homozygous parent. The resistance of melons to mildew is an example of this; several spots of mildew may be visible on the heterozygous F1 commercial varieties. Sometimes it is necessary to apply complementary anti-mildew treatments.

The heredity of resistance can be polygenic. Commercial varieties may include only one or two of the 'n' genes which are necessary to ensure maximum resistance. These varieties are likely to have different resistance levels, and are described as having the partial resistance or lower susceptibility. An example of this is the resistance of melons to the 1–2 strain of *Fusarium oxysporum* f.sp. *melonis*, or of cucumbers to mildew or Cucumber Mosaic Virus.

The development of pathogens (appearance of different strains capable of overcoming the resistant genes and making them ineffective) should not be confused with the effects of other pathogens causing identical symptoms. A major example of this is the two fungi which cause mildew in cucurbits: *Sphaerotheca fuliginea* and *Erysiphe cichoracearum*; the symptoms are exactly the same, but in the field the populations of these pathogens vary. For example, outbreaks of *E. cichoracearum* occur in spring and may gradually be replaced with *S. fuliginea*. In this case, a variety resistant to *E. cichoracearum* and susceptible to *S. fuliginea* will be resistant at the beginning of the season but will later become susceptible. It should not be concluded from this that the plant has lost its resistance to mildew. In the same way, a fungicide which is effective on *E. cichoracearum* will be less so when *S. fuliginea* takes over.

Some resistant characteristics introduced or in course of introduction into cultivated varieties of the cucumbers.

Pathogenic agent	Heredity	Level[*]
Cucumber Mosaic Virus	polygenic	variable
Papaya Ring Spot Virus	monogenic (wmv-1-1)	high
Zucchini Yellow Mosaic Virus	monogenic (zym)	high
Pseudomonas lachrymans	polygenic	variable
Cladosporium cucumerinum	monogenic (Ccu)	high
Corynespora cassiicola	monogenic (Cca)	high
Sphaerotheca fuliginea	polygenic	variable
Erysiphe cichoracearum	polygenic	variable
Pseudoperonospora cubensis	polygenic	variable

[*] Resistance level varies according to the variety. Some cucumbers grown elsewhere have resistance to diseases not known or almost unknown in France, such as bacterial wilt (*Erwinia tracheiphila*), or fusarium wilt (*Fusarium oxysporum* f.sp. *cucumerinum*).

Some resistant characteristics introduced or in course of introduction into cultivated varieties of melon in France.

Pathogenic agent	Heredity	Level[*]
Cucumber Mosaic Virus	polygenic	variable
Papaya Ring Spot Virus	monogenic (Prv)	very high
Zucchini Yellow Mosaic Virus	monogenic (Zym)	high
Transmission of all viruses by *Aphis gossypii*	monogenic (Vat)	high
Melon Necrotic Spot Virus	monogenic (nsv)	very high
Sphaerotheca fuliginea	monogenic	high
Erysiphe cichoracearum	monogenic	high
F. oxysporum f.sp. *melonis* strains 0 and 2	monogenic (Fom-1)	high
F. oxysporum f.sp. *melonis* strains 0 and 1	monogenic (Fom-2)	high
F. oxysporum f.sp. *melonis* strain 1–2	polygenic	variable

[*] Level of resistance varies according to the variety.

MELON

Abbreviations used for disease resistance in the melon (the presence of the Fn gene has been indicated because although it does not affect resistance it is useful for diagnosis)

F0 = *Fusarium oxysporum* f.sp. *melonis* strains 1 and 2 (Fom-1 gene)
F1 = *Fusarium oxysporum* f.sp. *melonis* strains 0 and 1 (Fom-2 gene)
F1-2 = *Fusarium oxysporum* f.sp. *melonis* strains 1 and 2 (polygenic)
Sf1 = *Sphaerotheca fuliginea* strain 1
Sf2 = *Sphaerotheca fuliginea* strains 1 and 2
Ec = *Erysiphe cichoracearum*
Ag = *Aphis gossypii*
Fn = Wilting and necrosis with some strains of Zucchini Yellow Mosaic Virus (ZYMV)

Smooth Charentais type

68-02	F0, F1, Sf1, Sf2, Ec
Acor	F0, F1, Fn
Alpha	F0, F1, Fn
Athos	F0, F1
Bastion	F0, F1, Sf1, Sf2, Ec
Cantalun	F0
Cantor	F0, F1, Fn
Carlo	F0, F1, Sf1, Sf2, Ec, Fn
Charmel	F0, F1, F1-2, Sf1, Sf2
Costade	F0, F1, Fn
Cristel	F0
Cristo	F0, F1, Sf1, Sf2, Ec
Delta	F0, F1, Fn
Diamex	F0, Ec
Domus	F0, F1
Doublon	F0, Fn
Fusano	F1, Sf1, Sf2, Ec
Galoubet	F0, F1
Garrigue	F0, F1, Ec, Fn
Glanum	F1, Sf1, Sf2, Ec
Hermes	F0, F1
Ido	F0, Sf1
Jador	F0, F1, F1-2, Sf1, Sf2, Fn
Jerac	F0, F1, Fn
Jet	F0, F1, Ec, Fn
Jivaro	F0, F1
Laro	F0, F1, Fn
Luberon	F0, F1
Maestro	F0, F1, Sf1, Sf2, Ec
Margot	F0, F1, Ag
Orlinabel	F0
Paradou	F0, F1
Pharo	F0, F1, Sf1, Sf2, Ec
Preco	F0, F1
Presto	F0, F1, Sf1, Sf2, Ec, Fn
Prior	F0, F1, Sf1, Sf2, Ec
Santon	F1
Savor	F0, F1, Ec
Soldor	F0, F1, F1-2, Sf1, Sf2
Tabor	F0, F1
Talma	F0, F1, Fn
Troubadour	F0, F1
Vedrantais	F0
Viva	F0, F1

Netted Charentais type

Accent	F0, F1, Sf1, Sf2, Ec
Alienor	F0, F1, Fn
Bolero	F0, F1, Sf1, Sf2, Ec, Fn
Comet	Sf1, Sf2, Ec
Concorde	F0, F1, Sf1, Sf2, Ec, Fn
Fiesta	F0, F1, Sf1, Sf2, Ec
Gama	F0, F1, Fn
Haros	F0, F1
Mab	F0, F1, Fn
Orus	Fo, F1
Oscar	F0, F1, Fn
Pallium	F0, F1, Sf1, Sf2, Ec
Pancha	F0, Sf1, Sf2, Ec
Panchito	F0, F1, Sf1, Sf2, Ec
Ramon	F0, F1
Rasto	F0, F1, Sf1, Sf2, Ec
Romeo	F0, Sf1, Fn
Sierra	F0, F1, Sf1, Sf2, Ec
Sprint	F0, Fn (Disj)
Sucdor	F0, F1, Sf1, Sf2, Ec

Elongated netted type

Bredor	F1, Fn
Calipso	F0, F1, Sf1, Sf2, Ec, Fn
Dogo	F1, Fn
Elton	F0, F1, Sf1, Sf2, Ec
Euromarket	F0, Ec
Fiata	F0, F1, Sf1, Sf2, Ec
Fox	F0, F1, Fn
Lothar	F0, F1, Sf1, Sf2, Ec
Mambo	F0, F1, Fn, Ag
Pacio	F0, F1
Parsifal	F0, F1, Sf1, Sf2, Ec
Rekord	F0, F1, Fn
Retor	F0, F1, Fn
Soleado	F0, F1, Sf1, Sf2, Ec
Supermarket	Sf1, Sf2, Ec

Round netted type

Durango	F0, Sf1, Sf2, Ec
Fastoso	F0, F1, Fn
Hymark	Sf1, Sf2, Ec
Topscore	Sf1, Sf2, Ec

Canary type	
Amber	F1, Sf1, Sf2, Ec, Fn
Aril	F1, Ec
Canador	F1, Ec, Fn
Doral	F0, Sf1, Sf2, Ec
Eloro	F1, Sf1, Sf2, Ec
Helios	F1, Sf1, Ec, Fn
Lutina	F1, Ec, Fn
Pandor	F1, Fn
Sirocco	F1, Ec

Winter Olive type	
Albor	F0, F1, Sf1, Ec, Fn
Caramello	F0, F1

Elisap	F1, Sf1, Ec, Fn
Noble	Sf1, Sf2, Ec
Rodos	F1, Ec, Fn

Other types	
Dikti	F1, Fn
Galia	Sf1
Gallicum	Sf1
M78.01	F0, F1, Fn
M78.02	F0, Sf1, Sf2, Ec, Fn
Pamir	F0, F1, Ec, Fn
Polidor	F0
Regal	F1, Sf1, Sf2, Ec
Silando	F0, F1

CUCUMBER

Abbreviations used for disease resistance in cucumbers:

Cl = *Cladosporium cucumerinum*
Co = *Corynespora cassiicola*
CMV = Cucumber Mosaic Virus
O = Oïdium (*Sphaerotheca fuliginea* and/or *Erysiphe cichoracearum*)
M = *Pseudoperonospora cubensis* (Mildew)
P = *Pseudomonas lachrymans* (Angular spots)
PRSV = Papaya Ring Spot Virus (formerly WMV1)
ZYMV = Zucchini Yellow Mosaic Virus

Bet Alpha or Mini Type	
Akhdar	C1, O
Arabel	CMV
Arabio	Cl, O
Astarte	Cl, Co, CMV
Bahia	CMV
Banza	Cl, CMV, O
Bet Alpha	CMV
Cantor	CMV
Caprice	Cl, CMV
Carlton	M
Comet A	Cl, O, P
Cora	CMV
Cordito	Cl, O
Dina	Cl
Douceur	CMV, PRSV
Emily	Cl, CMV
Esmeralda	Cl
Farid	Cl, O, M
Farol	Cl, O
Figaro	Cl, O
Hamada	CMV, O, M
Jabal	Cl, O, M
Jamil	Cl, O, M
Karim	Cl, O, M
Khalifa	CMV
Maram	Cl
Meta	Cl, CMV, O
Midistar	O, M
Minisol	Cl
Miracross	CMV, PRSV
Mirella	CMV
Monarch	Cl, CMV, O, M, P

Nabil	Cl, O, M
Noor	Cl, O
Pamfilia	Cl, Co, CMV
Paska	Cl, CMV, O
Petita	Cl, Co, CMV
Picobello	Cl
Pigal	O, M
Ramita	CMV, O, M
Rawa	Cl, CMV, O, M
Reem	O, M
Sabeel	Cl, O
Safaa	CMV, O, M
Sahara	Cl
Samar	Cl, Co, O
Silor	Cl, CMV, O
Taha	Cl, O, M
WSM 55	Cl, O
WSM 64	Cl, O
Zena	CMV, PRSV

Gherkin type	
Accordia	Cl, CMV, O
Adonis	Cl, CMV, M
Alert	Cl, CMV, M
Alvira	Cl, CMV, O
Amanda	Cl, CMV, O
Anka	Cl, CMV, O
Anuschka	Cl, CMV, O, M
Arena	Cl, CMV, M
Bix Domino	Cl, CMV, O, M
Bounty	Cl, CMV, O, M, P
Burgos	Cl, CMV, O, M

Calypso	Cl, CMV, O, M, P	Gynial	Cl, CMV
Capir	Cl, CMV, O, M	Highmark	Cl, CMV, M
Carduro	Cl, CMV, O	Jazzer	Cl, CMV, O, M
Carolina	Cl, CMV, O, M, P	Le Généreux	CMV
Carpadon	Cl, CMV, O	Marenka	Cl, CMV, O, M, P
Ceto	Cl, CMV	Marketmore	P
Christine	Cl, CMV, O	Marketmore 76	Cl, CMV, O, M
Colet	Cl, CMV, O	Monarch	Cl, CMV, O, M, P
Davista	Cl, CMV	N° 70	Cl, CMV, O, P
Donja	Cl, CMV, O, M	Poinsett	O
Doplus	Cl, CMV, O	Poinsett 76	Cl, O, M, P
Dura	Cl, CMV, O	Prestige	CMV
Elena	Cl, CMV, O	Slice King	O, M, P
Elon	Cl, CMV, O	Sprint	Cl, CMV, O
Fablo	Cl, CMV	Sprint 440	Cl, CMV, O, M, P
Fancipak	Cl, CMV, O, M, P	Striker	Cl, CMV, O, M, P
Fanfare	Cl, CMV, O	Sweet Slice	CMV, O, M, PRSV, ZYMV
Flurry	Cl, CMV, O, M, P	Victory	Cl, CMV, O, M
Fortos	Cl, CMV, O		
Ginor	Cl, CMV		
Hyclos	CMV	**Glasshouse Cucumber Type**	
Ilonca	Cl, CMV, O	Amazone	Cl
Inge	Cl, CMV, O	Aramon	Cl, O
Kobus	Cl, CMV, O, M, P	Ardo	CMV
Leonore	Cl, CMV, O	Aurelia	Cl, Co
Levina	Cl, CMV, O	Avir	Cl
Levo	Cl, CMV	Bambina	Cl
Lucky Strike	CMV, M	Belitta	Cl, Co
Marinda	Cl, CMV, O	Bella	Cl, Co, O
Maxor	Cl, CMV	Birgit	Cl, Co
Melani	Cl, CMV, O	Boloria	Cl, Co
Meresto	Cl, CMV, O	Boneva	Cl, Co
Metula	Cl, CMV, O	Brudania	Cl, Co
Milex	Cl, CMV, O	Brunex	Cl, Co
Milglas	Cl, CMV, O	C 445	Cl
Naf Fanto	Cl, CMV, O	C 561	Cl
Ouverture	Cl, CMV, O	Camirex	Cl, Co, O
Parifin	Cl, CMV	Canex	Cl, Co, O, M
Parigyno	CMV	Cargo	M, P
Parker	Cl, CMV, O	Carmen	O, M
Parmel	Cl, CMV, O	Cilla	Cl, Co
Parnita	CMV, O, M	Colias	Cl, Co
Paula	Cl, CMV, O	Cordoba	Cl, O
Pepito	Cl, CMV, O	Corona	Cl, Co
Pik Rite	CMV, O, M, P	Daleva	Cl
Pionnier	Cl, CMV, O	Dalibor	Cl
Premier	Cl, CMV, O, M, P	Dorian	Cl, Co
Profi	Cl, CMV, O	Elka	Cl, Co
Score	Cl, CMV, O	Euphya	O
Sena	Cl, CMV, M	Falko	Cl
Talgo	Cl, CMV, O	Famosa	Cl
Tamor	Cl, CMV, O, M, P	Farbio	Cl
Tomara	Cl, CMV, O	Farbiola	Cl
Triplecrown	CMV, O, P	Farona	Cl
Wilma	Cl, CMV, O	Fembasy	Cl, Co, O, M,
Witlo	CMV, O	Fertila	Cl
		Fidelio	Cl, O, M
Spiny Cucumber Type		Fytos	Cl, O
Amiral	Cl, CMV, O	Gador	Cl, Co, O
Astrea	Cl, CMV, O	Girola	Cl, Co
Belair	O, M, P	Grandiosa	Cl
Belcanto	Cl, Co, CMV, O, P	Ingrid	Cl, Co
Bellando	Cl, CMV, O, M, P	Jessica	Cl
Breso	Cl, CMV, O	Kamaron	Cl, Co
Comet A	Cl, CMV, O, M, P	Kivia	Cl, Co
Darina	Cl, CMV, O, M, P	Lucinde	Cl

Manu	Co	Salvador	Cl
Marillo	Cl, O	Samba	Cl
Mildana	Cl, Co, O, M	Sanabel	Cl
Millio	Cl, O	Sandoro	Cl
Mustang	Cl, Co	Sandra	Cl, Co
Noval	Cl, M	Saskia	Cl, Co
Palmera	Cl	Sofia	Cl, Co
Pandorex	Cl, Co	Sortena	Cl
Pepinex	Cl, Co	Stereo	Cl
Perex	Cl, Co, O, M	Superator	Cl, Co
Primio	Cl	Toska 70	Cl
Profito	O	Valore	Cl
Radja	Cl, Co	Ventura	Cl
Rebella	Cl	Verana	Cl, Co
Regina	Cl	Vitalis	Cl
Rival	Cl, P		

INDEX

MICRO-ORGANISMS CITED

Micro-organisms responsible or not for parasitic diseases

Normal type = pathogen
Bold type = **factfiles** (biology, control)

Actinomycetes 113, 129

Bacteria
Agrobacterium rhizogenes 113, 129 (**factfile 4**)
Erwinia ananas 183 (**factfile 4**)
Erwinia aroïdea 183 (**factfile 4**)
Erwinia carnegieana 183 (**factfile 4**)
Erwinia carotovora var. *carotovora* 125, 131, 143, 183 (**factfile 4**)
Erwinia spp. 113, 117, 125, 183, 184
Erwinia tracheiphila 131, 143, 151 (**factfile 3**)
Pseudomonas marginalis 105
Pseudomonas viridiflava 105
Pseudomonas syringae pv. *lachrymans* 71, 75, 85, 97, 99, 153, 161, 276 (**factfile 1**)
Pseudomonas syringae (group II) 71, 75, 85
Xanthomonas campestris pv. *cucurbitae* 71, 75, 85, 99, 153, 161, 163, 183, 184, 187, 189 (**factfile 2**)
Xanthomonas campestris pv. *melonis* 183, 184

Fungi
Acremonium sp. 71, 103, 262
Alternaria alternata 71, 75, 87
Alternaria cucumerina 71, 75, 87, 89 (**factfile 12**)
Alternaria tenuis 87, 183, 184 (**factfile 13**)
Alternaria pluriseptata 71, 75, 87
Ampelomyces quisqualis 71, 103
Ascochyta citrullina see *Didymella bryoniae*
Athelia rolfsii see *Sclerotium rolfsii*
Botryotinia fuckeliana see *Botrytis cinerea*
Botrytis cinerea 71, 93, 131, 139, 141, 142, 153, 167, 169, 175, 176, 183, 184 (**factfile 7**)
Cephalosporium cucurbitarum 254
Cephalosporium roseum see *Trichothecium roseum*
Cephalosporium spp. 109
Cercospora citrullina 71, 75 (**factfile 12**)
Cercospora cucurbitae 71, 87
Cercospora melonis 71, 87 (**factfile 12**)
Chalara elegans 113, 125, 254
Choanephora cucurbitae 153, 175, 179, 183, 184 (**factfile 13**)
Cladosporium cucumerinum 71, 75, 131, 135, 153, 161, 163, 167, 187, 251, 276 (**factfile 5**)
Cladosporium spp. 71, 103, 153, 171, 183, 184 (**factfile 13**)
Colletotrichum lagenarium 71, 75, 131, 135, 153, 159, 161, 163, 167 (**factfile 6**)
Corticium rolfsii 125, 153, 171
Corticium solani 113, 125
Corynespora cassiicola 71, 75, 87, 89, 153, 276 (**factfile 12**)
Didymella bryoniae 71, 93, 95, 113, 117, 119, 127, 131, 132, 139, 141, 142, 153, 171, 175, 176, 183 (**factfile 8**)
Diplodia gossypina see *Diplodia natalensis*
Diplodia natalensis 153, 175, 177
Erysiphe cichoracearum 71, 101, 103, 131, 135, 153, 167, 169, 177 (**factfile 9**)
Fulago septica 129
Fusarium equiseti 254
Fusarium javanicum 127
Fusarium moniliforme var. *subglutinans* 109
Fusarium oxysporum f.sp. *benincasae* 151
Fusarium oxysporum f.sp. *cucumerinum* 131, 143, 145, 151 (**factfile 22**)
Fusarium oxysporum f.sp. *lagenariae* 151
Fusarium oxysporum f.sp. *melonis* 37, 41, 53, 55, 131, 143, 145, 150, 153, 175, 251, 254, 276 (**factfile 23**)
Fusarium oxysporum f.sp. *momordicae* 151

Fusarium oxysporum f.sp. *niveum* 131, 143, 145, 151 (**factfile 24**)
Fusarium roseum 183, 184
Fusarium solani 109, 184
Fusarium solani f.sp. *cucurbitae* 127, 183, 254 (**factfile 18**)
Fusarium spp. 113, 117, 119, 127, 131, 139, 141, 153, 175, 183, 184
Geotrichum candidum 183, 184 (**factfile 13**)
Glomerella cingulata var. *orbiculare* see *Colletotrichum lagenarium*
Helminthosporium cucumerinum 71, 75, 89
Leandria momordica 71, 75, 89
Leveillula taurica 71, 101 (**factfile 10**)
Macrophomina phaseoli 113, 125, 131, 135, 139, 153, 171 (**factfile 13**)
Microdochium tabacinum see *Monographella cucumerina*
Monographella cucumerina 113
Monosporascus cannonbalus 113
Monosporascus eutypoïdes 113 ·
Mucor spp. 153, 183
Mycosphaerella citrullina see *Didymella bryoniae*
Mycosphaerella melonis see *Didymella bryoniae*
Myrothecium roridum 71, 75, 89, 113, 153, 187
Oïdiopsis taurica see *Leveillula taurica*
Olpidium spp. 75, 81, 117, 139
Oospora lactis see *Geotrichum candidum*
Penicillium crustosum see *Penicillium oxalicum*
Penicillium digitatum 153, 175, 176
Penicillium oxalicum 113, 119, 127, 131, 139, 141 (**factfile 19**)
Penicillium spp. 71, 103, 153, 171, 183, 184
Phoma cucurbitarum see *Didymella bryoniae*
Phomopsis sclerotioïdes 113, 117, 119, 125 (**factfile 19**)
Physalospora rhodina see *Diplodia natalensis*
Phytophthora capsici 71, 75, 121, 131, 135, 139, 153, 171, 175, 177, 181, 183, 184 (**factfile 14**)
Phytophthora cryptogea 121 (**factfile 14**)
Phytophthora drechsleri 121 (**factfile 14**)
Phytophthora melonis 121
Phytophthora nicotianae 177
Phytophthora sinensis 121
Phytophthora spp. 113, 117, 120, 121, 153, 171, 183 (**factfile 14**)
Pleospora herbarum see *Stemphylium botryosum*
Pseudoperonospora cubensis 37, 41, 53, 71, 93, 95, 97, 101, 105, 109, 276 (**factfile 11**)
Pyrenochaeta lycopersici 113, 117, 118, 123 (**factfile 17**)
Pythium spp. 113, 117, 118, 121, 153, 171, 177, 181, 183, 254, 255 (**factfile 14**)
Pythium aphanidermatum 121, 177, 181, 183 (**factfile 14**)
Pythium butleri see *Pythium aphanidermatum*
Pythium debaryanum see *Pythium irregulare*
Pythium intermedium 121 (**factfile 14**)
Pythium irregulare 121, 177, 181 (**factfile 14**)
Pythium splendens 121 (**factfile 14**)
Pythium ultimum 121, 177, 183 (**factfile 14**)
Rhizoctonia solani 109, 113, 117, 118, 123, 127, 131, 135, 153, 171, 254, 255 (**factfile 15**)
Rhizopus nigricans 153, 171, 175, 179, 183, 184 (**factfile 13**)
Rhizopus stolonifer see *Rhizopus nigricans*
Sclerotinia sclerotiorum 131, 132, 139, 141, 153, 175, 176, 183, 184 (**factfile 20**)
Sclerotium rolfsii 125, 153, 171
Septoria cucurbitacearum 71, 75 (**factfile 12**)
Sphaerothecae fuliginea 71, 101, 103, 131, 135, 153, 167, 169, 276, 277 (**factfile 9**)
Stemphylium cucurbitacearum 89
Stemphylium botryosum 184
Stephanoascus sp. 71, 103
Thanatephorus cucumeris see *Rhizoctonia solani*
Thielaviopsis basicola see *Chalara elegans*
Tilletiopsis spp. 71, 103
Trichothecium roseum 153, 175, 176, 183, 184
Ulocladium atrum 71, 75, 87, 99 (**factfile 12**)
Ulocladium botrytis 184
Ulocladium consortiale 87
Ulocladium cucurbitae see *Ulocladium atrum*
Verticillium albo-atrum see *Verticillium dahliae*

Verticillium dahliae 131, 143, 145, 149 (**factfile 25**)
Verticillium lecanii 71, 103

Viruses and Viroids
Melon luteovirus 37, 55, 57 (**factfile 33**)
Cucumber Pale Fruit Viroid (CPFV) 191 (**factfile 35**)
Melon Necrotic Spot Virus (MNSV) 71, 75, 82, 97, 131, 139, 276 (**factfile 28**)
Cucumber Soil Borne Virus (CSBV) (**factfile 28**)
Cucumber Yellows Virus (CYV) 37, 43, 55 (**factfile 28**)
Muskmelon Yellows Virus (MYV) 37, 43, 55, 57 (**factfile 33**)
Lettuce Infectious Yellows Virus (LIYV) 37, 55, 57 (**factfile 33**)
Cucumber Necrosis Virus (CNV) 71, 75, 82 (**factfile 28**)
Tobacco Necrosis Virus (TNV) 71, 75, 82 (**factfile 28**)
Cucumber Yellow Vein Virus (CYVV) 57 (**factfile 33**)
Cucumber Green Mottle Mosaic Virus (CGMMV) 27, 31, 37, 41, 43, 49 (**factfile 26**)
Cucumber Mosaic Virus (CMV) 19, 25, 27, 29, 31, 33, 37, 41, 43, 45, 47, 49, 51, 55, 105, 112, 153, 159, 191,
    193, 195, 276 (**factfile 29**)
Squash Mosaic Virus (SqMV) 19, 27, 31, 37, 41, 43, 45, 49, 51, 193 (**factfile 27**)
Zucchini Yellow Mosaic Virus (ZYMV) 19, 25, 27, 29, 31, 33, 37, 41, 43, 45, 49, 51, 55, 57, 105, 112, 135, 137,
    153, 186, 187, 191, 193, 195, 276 (**factfile 32**)
Water Melon Mosaic Virus type 1 *see* PRSV
Water Melon Mosaic Virus type 2 (WMV2) 19, 25, 27, 31, 33, 37, 41, 43, 45, 47, 49, 51, 153, 191, 193, 276
    (**factfile 31**)
Cucumber Toad Skin Virus (CTSV) 19, 25, 27, 29, 37, 41, 43, 49, 55, 153, 191, 193, 195 (**factfile 34**)
Beet Pseudo-Yellow Virus (BPYV) (**factfile 33**)
Papaya Ring Spot Virus (PRSV) 19, 27, 31, 33, 37, 41, 43, 45, 49, 51, 69, 153, 193, 195, 276 (**factfile 30**)
Cucumber Leaf Spot Virus (CLSV) (**factfile 28**)

DISEASES

Parasitic diseases

Bacterial diseases

Bacterial wilt 131, 143, 151 (**factfile 3**)
Angular leaf spots 71, 75, 85, 97, 99, 153, 161, 276 (**factfile 1**)
Bacterial rot of gourd fruit 71, 75, 85, 99, 153, 161, 163, 183, 184, 187, 189 (**factfile 2**)

Fungal diseases

Gummosis or 'grey' anthracnose 71, 75, 131, 135, 153, 167, 187, 251, 276 (**factfile 5**)
'Red' anthracnose 71, 75, 131, 135, 153, 159, 161, 163, 167 (**factfile 6**)
Grey mould 71, 93, 131, 139, 141, 142, 153, 167, 169, 175, 176, 183, 184 (**factfile 7**)
Sticky cankers caused by *Didymella* 71, 93, 95, 113, 117, 119, 127, 131, 132, 139, 141, 142, 153, 171, 175, 176, 183 (**factfile 8**)
Oïdium (powdery mildew) 71, 101, 103, 131, 135, 153, 167, 169, 277 (**factfile 9**)
Downy mildew 37, 41, 53, 71, 93, 95, 97, 101, 105, 109, 276 (**factfile 11**)
'Shot-hole' leaf spots 71, 75, 87, 89, 153, 276 (**factfile 12**)
Alternariosis (early blight) 71, 75, 87, 89 (**factfile 12**)
Cercospora leaf spot 71, 75 (**factfile 12**)
Leaf spot 71, 75 (**factfile 12**)
Various fruit rots 153, 171, 175, 179, 183, 184 (**factfile 13**)
Damping-off and root rot 113, 117, 118, 121, 249 (**factfile 14**)
Root deterioration due to *Rhizoctonia* 109, 113, 117, 118, 123, 127, 131, 135, 153, 171, 254, 255 (**factfile 15**)
Black rot of roots 113, 117, 119, 125 (**factfile 16**)
Corky root disease 113, 117, 118, 123 (**factfile 17**)
Collar fusarium wilt 127, 183, 254 (**factfile 18**)
'Blue' cankers due to *Penicillium* 113, 119, 127, 131, 139, 141 (**factfile 19**)
Cankers due to *Sclerotinia* 131, 132, 139, 141, 153, 175, 176, 183, 184 (**factfile 20**)
Vascular fusarium wilt of cucumbers 131, 143, 145, 151 (**factfile 22**)
Vascular fusarium wilt of melon 37, 41, 53, 55, 131, 143, 145, 150, 153, 175, 251, 254, 276 (**factfile 23**)
Vascular fusarium wilt of water melon 131, 143, 145, 151 (**factfile 24**)
Verticillium wilt 131, 143, 145, 149 (**factfile 25**)

Virus diseases

Cucumber mottling 27, 31, 37, 41, 43, 49 (**factfile 26**)
Melon Necrotic Spot Disease 71, 75, 82, 97, 131, 139, 276 (**factfile 28**)
Cucumber Mosaic Disease 19, 25, 27, 29, 31, 33, 37, 41, 43, 45, 47, 49, 51, 55, 105, 112, 153, 159, 193, 195, 276 (**factfile 29**)
Papaya Ring Spot Disease 19, 27, 31, 33, 37, 41, 43, 45, 49, 51, 69, 159, 193, 276 (**factfile 30**)
Water Melon Mosaic Disease 19, 25, 27, 31, 33, 37, 41, 43, 45, 47, 49, 51, 153, 191, 193, 276 (**factfile 31**)
Zucchini Yellow Mosaic Disease 19, 25, 27, 29, 31, 33, 37, 41, 43, 45, 49, 51, 55, 57, 105, 112, 135, 137, 153, 186, 187, 191, 193, 276 (**factfile 32**)

Non-parasitic or physiological diseases
Adverse climatic conditions and poor cultivation operations 19, 25, 27, 29, 71, 91, 105, 111, 249, 254
Genetic abnormalities or characteristics (chimera etc.) 37, 67, 97, 99, 153, 191, 192, 249
Stylar abnormalities in cucumber fruit 153, 175, 177
Production of male flowers in gynoecious cucumber varieties 153, 157
Silvering in courgettes 37, 69
Root asphyxiation 113, 117
Blossom-end rot 153
Cluster of apical flowers 153, 157
Nutritional deficiencies 19, 25, 55, 59, 60, 61, 153, 195, 197
Corky stylar retraction 153, 175, 177
Fruit drop 153, 157, 159, 175
Sun scald 153, 171, 173
Crown blight 105, 111, 276
Faulty fruit set 153, 175
Hail damage 131, 135, 137, 153, 167, 169
Pollen deposits 71, 103, 104, 167, 169
Physiological splitting 113, 129

PESTS AND PARASITIC PLANTS

PHOTOGRAPHS OF SYMPTOMS CAUSED BY MICRO-ORGANISMS

(micro-organisms responsible or not for parasitic diseases)

Bacteria

Fungi

PHOTOGRAPHS OF DISEASE SYMPTOMS

Bacterial diseases

Fungal diseases

Gummy cankers caused by *Didymella* 116, 163, 164, 169, 171, 172, 246, 247, 269, 272, 278, 346, 347, 348, 363, 364, 366, 369

Oidium (powdery mildew) 111, 186, 187, 188, 189, 193, 194, 195, 196, 197, 198, 199, 261, 338

Mildew 3, 74, 110, 173, 174, 175, 176, 177, 178, 179, 180, 181, 182, 190, 209

Damping-off and root rot 208, 223, 226, 227, 230, 231, 232, 233, 359, 419, 420, 421

Root deterioration due to *Rhizoctonia* 224, 234, 235, 236, 237, 263, 344

Black rot of roots 201, 204, 220, 221, 228, 229, 240, 241, 242, 243, 244

Corky root disease 203, 238

'Blue' cankers due to *Penicillium* 248, 273

Cankers due to *Sclerotinia* 267, 268, 274, 275, 355, 365

Vascular fusarium wilt of cucumbers 257, 281, 284, 294, 295, 296, 297

Vascular fusarium wilt of melon 1, 73, 98, 205, 282, 283, 285, 286, 287, 288, 289, 358, 422

Vascular fusarium wilt of water melon 301, 302, 303

Verticillium wilt 280, 290, 291, 292, 293, 298, 299, 300

Virus diseases

Mottling of Cucumber 24, 62

Melon Necrotic Spot Disease 138, 139, 140, 141, 142, 143, 183, 270

Cucumber Mosaic Disease 4, 5, 11, 13, 14, 19, 22, 45, 46, 47, 51, 53, 54, 55, 56, 61, 65, 84, 202, 218, 306, 389, 394, 398, 405, 412

Papaya Ring Spot Disease 18, 26, 33, 49, 68, 104, 405

Water Melon Mosaic Disease 19, 29, 32, 55, 57, 66, 390

Zucchini Yellow Mosaic Disease 6, 7, 8, 9, 12, 20, 21, 23, 27, 28, 29, 30, 31, 34, 35, 41, 48, 50, 52, 56, 58, 63, 67, 69, 70, 71, 72, 83, 219, 264, 307, 316, 381, 382, 391, 392, 395, 399, 400, 401, 402, 403, 404, 405

Non-parasitic or physiological diseases

Adverse climatic conditions and poor cultivation operations 10, 16, 25, 162, 213, 214, 215, 239, 251, 266

Genetic abnormalities or characteristics (chimera etc.) 44, 96, 97, 100, 101, 102, 103, 105, 184, 185, 265, 397

Stylar abnormalities in cucumber fruit 361

Production of male flowers in gynoecious cucumber varieties 310

Nutritional deficiencies 17, 86, 87, 88, 89, 410

Corky stylar retraction 360

Fruit drop 313, 314, 315

Sunburn 349, 350, 351, 352

Crown blight 216, 217

Faulty fruit set 313, 314, 315

Hail damage 266, 340, 341

Pollen deposits 200, 342

Physiological splitting of the stem 252

Physiological splitting of the fruit 386

'Strangulation' of cucumber fruit 409

Excess of male flowers in the melon 309

Excess salinity 94

Stem fasciation 265

Growth cracks 386

Fasciated fruit 407, 408

Epidermal splits in cucumber fruit 385, 387

Physiological drying in melons 210, 211

Hypersusceptibility to mildew 97

Oedemas 191, 192, 327, 328, 331

Yellowing of melon fruit 393

Physiological yellowing of courgette leaves 96

Peripheral necrosis of the lamina 95

Fasciated nodes in cucumbers 312

Root regression in cucumbers 208, 225

Various phytotoxicities 2, 36, 37, 38, 39, 40, 42, 90, 91, 92, 93, 94, 95, 99, 106, 107, 159, 160, 161, 207, 212, 253, 329, 330, 353, 413, 414, 415, 416, 417, 418

Giant sepals on cucumber flowers 311

Glassiness in melon fruit 377, 378

PHOTOGRAPHS OF DAMAGE BY PESTS AND PARASITIC PLANTS

PHOTOGRAPHS OF SOME OF THE CUCURBITACEAE BELONGING TO THE GENERA

Main Works of Reference

- *Diseases of greenhouse plants*, 1984, J.T. Fletcher; Longman Inc., New York, USA.
- *Identifying Diseases of Vegetables*, 1983, A.A. MacNab, A.F. Sherf, J.K. Springer; Pennsylvania State University, USA.
- *Les maladies des plantes maraîchères* (Diseases of market-garden plants), 1970, C.-M. Messiaen, R. Lafon; INRA Publication 6-70, France.
- *Cucurbit Diseases, A practical guide for seedsmen, growers and agricultural advisers*, 1988, E. Bernhardt, J. Dodson, J. Waterson; Petoseed Co. Inc., USA.
- *Vegetable Diseases and their Control*, 1986, A.F. Sherf, A.A. MacNab; John Wiley & Sons, Bristol, UK.
- *Phytopathologie des pays chauds* (Plant pathology in hot countries), 1951, L. Roger, Paul Lechevalier, Editor, France.
- *Le potager tropical* (The tropical vegetable garden), 1989, C.-M. Messiaen; 2nd ed. Presses Universitaires de France, France.